Poisson Process and its Fractional Extensions with Applications

This book accompanies the reader from the simple, standard Poisson process to its more complex transformations, using a unified framework, showing proofs of basic theorems and references to more difficult results.

The reader will also benefit from the empirical applications of the various models to real data in finance and insurance (e.g., stock market crashes and fire accidents), physics and geology (e.g., earthquakes), biology and demography (e.g., births, deaths, pandemics), and more. Many of the applications relate to financial mathematics, and a professional can glean a lot from an understanding of how to apply this important mathematical tool.

Some examples are worked out in open-source R showing the reader how to implement the models. Strengths of *Poisson Process and its Fractional Extensions with Applications* are:

- A focus on a very fundamental class of stochastic processes
- An introductory approach assuming no previous experience with the topics
- A wide range of generalizations considered, including recent advances to which one of the authors has made seminal contributions
- Attention not only to theory, but also to practical applications with real-data examples from a wide range of fields

The authors pay special attention not only to the theoretical foundations of the Poisson process and the various generalizations that have been proposed in the literature in the last decades, but also to the practical applications in many different fields of the Poisson models for a better understanding of the stochastic structure of the real world.

Note that for the sake of the reader, the probability generating functions are summarized in Appendix A; subordinators in Appendix B; fractional derivatives in Appendix C; gamma and Mittag-Leffler functions in Appendix D, and tools for data analysis in Appendix E. Appendix F contains the R code.

Chapman & Hall/CRC Financial Mathematics Series

Series Editors

Foundations of Quantitative Finance
Book IV: Distribution Functions and Expectations
Robert R. Reitano

Foundations of Quantitative Finance
Book V: General Measure and Integration Theory
Robert R. Reitano

Computational Methods in Finance, Second Edition
Ali Hirsa

Interest Rate Modeling
Theory and Practice, Third Edition
Lixin Wu

Data Science and Risk Analytics in Finance and Insurance
Tze Leung Lai and Haipeng Xing

Foundations of Quantitative Finance
Book VI: Densities, Transformed Distributions and Limit
Robert R. Reitano

Malliavin Calculus in Finance
Theory and Practice, Second Edition
Elisa Alòs and David Garcia Lorite

Arbitrage and Rational Decisions
Robert Nau

Equity Release Products
Radu S. Tunaru and Enoch B. Quaye

Quantitative Finance with Case Studies in Python
A Practical Guide to Investment Management, Trading and Financial Engineering
Chris Kelliher

XVA Analysis
Probabilistic, Risk Measure, and Machine Learning Issues
Stéphane Crépey

Risk Analysis in Finance and Insurance
Alexander Melnikov

Handbook of Sustainable Mathematical Finance
Peter Tankov and Ruixun Zhang

Foundations of Quantitative Finance
Book VII: Brownian Motion and Other Stochastic Processes
Robert R. Reitano

Poisson Process and its Fractional Extensions with Applications
Enzo Orsingher, Riccardo Cesari, and Vieri Mosco

For more information about this series please visit: https://www.routledge.com/Chapman-and-HallCRC-Financial-Mathematics-Series/book-series/CHFINANCMTH

Poisson Process and its Fractional Extensions with Applications

Enzo Orsingher, Riccardo Cesari
and Vieri Mosco

CRC Press
Taylor & Francis Group
Boca Raton London New York

CRC Press is an imprint of the
Taylor & Francis Group, an **informa** business
A CHAPMAN & HALL BOOK

First edition published 2026
by CRC Press
2385 NW Executive Center Drive, Suite 320, Boca Raton FL 33431

and by CRC Press
4 Park Square, Milton Park, Abingdon, Oxon, OX14 4RN

CRC Press is an imprint of Taylor & Francis Group, LLC

ISBN: 978-1-032-49605-4 (hbk)
ISBN: 978-1-032-50338-7 (pbk)
ISBN: 978-1-003-39801-1 (ebk)

DOI: 10.1201/9781003398011

Typeset in CMR12 font
by KnowledgeWorks Global Ltd.

Publisher's note: This book has been prepared from camera-ready copy provided by the authors.

Contents

Symbols

Symbol	Description
a.s.	almost surely
CDF	Cumulative Distribution Function
e.g.	*exempli gratia*
i.e.	*id est*
i.i.d.	independent identically distributed
m.g.f.	moment generating function
p.g.f.	probability generating function
PMF	Probability Mass Function
PP	probability-probability (plot)
QQ	quantile-quantile (plot)
r.v.	random variable
s.t.	*such that*

Preface

When Siméon-Denis Poisson (1781-1840) analysed in 1837 the limit of the binomial distribution obtaining the result now identified as the Poisson distribution, he certainly was not aware of the many situations in which his result could be proved useful. He obtained his distribution by considering the case of a large number of trials n and a small probability of an event $p_n = \frac{\lambda}{n}$, where λ (ω in Poisson's book) is constant and called the Poisson's intensity.[1]

It may be interesting to note that he applied his research "*sur la probabilité des jugements en matière criminelle et en matière civile*", with special attention, therefore, to social and human sciences. In this book, similarly, we paid a special attention not only to the theoretical foundations of the Poisson process and the various generalizations that have been proposed in the literature in the last decades but also to the practical applications, in many different fields, of the Poisson models, for a better understanding of the stochastic structure of the real world.

From the theoretical point of view, Ch. 1 is devoted to the homogeneous Poisson process $N(t)$ and its simplest transformations and generalizations.

One is the order-i-weighted Poisson, i.e. a weighted sum of i independent Poisson processes (par. 1.2): $\sum_{j=1}^{i} jN_j(t)$; another one is the Skellam process, which, instead, is the difference between two independent Poisson processes (par. 1.3): $N_1(t) - N_2(t)$.

[1]See Poisson (1837, par. 81, p. 205). It seems that this result had been previously obtained by De Moivre (1711, p. 219).

A different generalization is obtained by considering a time-variable intensity, $\lambda(t)$, generating the non-homogeneous Poisson process (par. 1.4).

The sum of a random number $N(t)$ of i.i.d. random variables X_j is called a compound Poisson process: $\sum_{j=1}^{N(t)} X_j$ and it is analyzed in par. 1.5.

The passage from time t to space S creates the Poisson random field $N(S)$, modelling random events in an arbitrary n-dimensional space (par. 1.6).

The literature has also proposed the time integral of the Poisson process (par. 1.7): $\int_0^t N(s)ds$ and the iterated Poisson, which is a Poisson process N_α sampled at Poisson random times N_β: $N_\alpha(N_\beta(t))$. This is equivalent in distribution to a compound Poisson with Poisson random variables: $\sum_{j=1}^{N_\beta(t)} N_j(t)$ (par. 1.8).

Finally, the notion of *shot noise* is considered in par. 1.9, which enables us to consider *Hawkes processes* $N^H(t)$, that are Poisson processes driven by a conditional intensity $\tilde{\lambda}(t)$ subject to *shot noise.*

Chapters 2, 3 and 4 have to do with special Poisson processes in which the intensities depend (linearly or non-linearly) on the previous number of cases k: λ_k. If the number of cases (population) is increasing, we have a birth process (ch. 2); if the number of cases is decreasing, we have a death process (ch. 3). The combination of the two is a birth-death process with different birth and death intensities, λ_k, μ_k (ch. 4). The theory of queues with random arrivals and departures in a service facility is essentially similar.

The final chapter (ch. 5) analyses some probabilistic extensions of the homogeneous Poisson process. In particular, as shown in ch.1, the state probabilities $p_k(t) \equiv \Pr\{N(t) = k\}$ satisfy the difference-differential equation:

$$\frac{dp_k(t)}{dt} = -\lambda(1-B)p_k(t) \tag{0.1}$$

where B is the space-shift operator, $Bp_k(t) = p_{k-1}(t)$.

The generalization, called space-fractional Poisson process (par. 5.1), is obtained from:

$$\frac{dp_k(t)}{dt} = -\{\lambda(1-B)\}^{\alpha} p_k(t) \quad 0 < \alpha < 1 \tag{0.2}$$

and an even more general space-fractional process (par. 5.2) comes from:

$$\frac{dp_k(t)}{dt} = -f(\lambda(1-B))p_k(t) \quad 0 < \alpha < 1 \tag{0.3}$$

where $f(x)$ is a Bernštein function. It can be shown that the resulting process $N^f(t)$ is a homogeneous Poisson process with a time change through a particular subordinator H^f: $N^f(t) = N(H^f(t))$.

Another generalization of the Poisson process is obtained using the probability generating function (p.g.f.):

$$G(u,t) \equiv \sum_{k=0}^{\infty} u^k p_k(t) \tag{0.4}$$

In fact, for the homogeneous Poisson process, the p.g.f. satisfies the partial differential equation (subject to the initial condition $G(u,0) = 1$):

$$\frac{\partial G(u,t)}{\partial t} = \lambda(u-1)G(u,t) \tag{0.5}$$

but if we substitute the (partial) derivative with the ν-order (Caputo) fractional derivative ($0 < \nu < 1$), we obtain the p.g.f. of a fractional Poisson process (par. 5.3) which is a ratio of Mittag-Leffler functions and whose state probabilities turn out to be a weighted average of homogeneous Poisson state probabilities.

In addition, the ν-order fractional derivative can be inserted into the difference-differential equation of the state probabilities (1), obtaining, as solution, the so-called time-fractional Poisson process (par. 5.4). The space-time fractional generalization (par. 5.5) combines the space-fractional and the time-fractional derivatives to generate a very general set of discrete-space, continuous-time processes of the Poisson family.

Finally applying the fractional derivative to the difference-differential equation of the standard (non-linear) birth-death process, you obtain the fractional counterpart (par. 5.6).

Note that, for the sake of the reader, the probability generating functions are summarized in Appendix A, subordinators in Appendix B, fractional derivatives in Appendix C, Gamma and Mittag-Leffler functions in Appendix D, and tools for data analysis in Appendix E.

As anticipated, in the spirit of the original Poisson's approach, the book is filled with a wide range of practical cases to which the Poisson models are applied. Starting from the famous horse-kick phenomenon used by Von Bortkewitsch (1898) as an early application of the Poisson distribution, we provide the reader with estimation methods and data analysis tools (see Appendix E) useful for the Poisson model implementation. The examples are taken from the real world and belong to the most disparate empirical fields in which point events and point processes are observed: sociology, games, medicine, finance, insurance, geology, epidemiology and biology. Different models are compared, and statistical tests are used to assess the models' fit. Many graphs are used to gain insights into the processes behaviour. The open source R code is sometimes provided to help the interested reader to analyze their own data and get familiar with the different Poisson models (See Appendix F).

About the Authors

Enzo Orsingher is a leading scholar with seminal contributions in the field of Stochastic Processes and Emeritus Professor of Probability at Sapienza University of Rome. His scientific interests cover many areas of probability and stochastics, especially the relation between heat-type equations and diffusions and the connection between wave equations, the telegraph process and random motions. He has explored other areas of probability such as Brownian motion in non-Euclidean spaces, pseudo-processes and fractional processes. Current interests are in random motions on the plane and in the three-dimensional space, with recent curiosities in the history of mathematics. He has written over 200 papers in peer-reviewed journals, visited several universities abroad and has supervised 17 PhD Students.

Riccardo Cesari is a full professor of mathematical methods of economics, finance and actuarial sciences at the University of Bologna. He has a DPhil from Oxford University. Between 2013 and 2025, he was a member of the board of IVASS, the Italian Supervisory Authority for Insurance Companies. His academic interests and publications concern the term structure of interest rates, valuation of structured securities, analysis of unit trusts and pension funds, forecasting models of financial markets and strategic and tactical optimal financial allocation, implicit guarantees in insurance contracts and solvency problems in life and non-life insurance.

Vieri Mosco is an actuarial researcher at IVASS. He has a PhD from Sapienza University of Rome, where he discussed a thesis on optimal management of life insurance segregated-funds. His current interest is in empirical applications of quantitative models to various aspects of motor insurance, including road security and fraud detection.

Part I

Poisson processes

1

Poisson Processes

CONTENTS

DOI: 10.1201/9781003398011-1

The first chapter covers the *homogeneous Poisson process* and its most common extension such as the *non-homogeneous Poisson process* through a classical introduction of its properties and alternative definitions. Other common processes related to the Poisson process such as *Compound Poisson processes* and the *Skellam process* (the difference of two Poisson processes) are introduced. As a first example of Poisson process with *multiple jumps*, the so-called *weighted Poisson process* (a weighted average of Poisson processes) is dealt with. A first example of *composition of Poisson processes* is provided by the *iterated Poisson process*, namely a Poisson processes whose time runs according (*subordinated*) to another Poisson process. Moreover, the *integral* (the area spanned below the path) *of Poisson processes* is dealt with also when the time increments are small. The shot noise process and Hawkes processes are also introduced. Finally, *Poisson random fields* are shortly mentioned.

1.1 The Homogeneous Poisson Process

A homogeneous Poisson process $N(t) = N((0,t]) = N(0,t]$, $t > 0$, is a continuous-time (one parameter, $\lambda > 0$) stochastic process taking integer values and representing the number of events happening over a *finite* time interval $(0,t]$. Its probabilistic evolution is governed by the following rules:

Definition 1.

1. $\Pr\{N(0) = 0\} = 1$;
2. $\Pr\{N(t, t + dt] = 1\} = \lambda dt + o(dt)$, with $\lambda > 0$;
3. $\Pr\{N(t, t + dt] = 0\} = 1 - \lambda dt + o(dt)$;
4. $\Pr\{N(t, t + dt] > 1\} = o(dt)$;
5. The process has independent increments in the sense that if $0 = t_0 < t_1 < \ldots < t_n < t$, the random variables $N(t_{j-1}, t_j] = N(t_j) - N(t_{j-1})$, $1 \leq j \leq n$, are independent.

By $N(s,t]$ we denote the number of events in $(s,t]$ with $s<t$. Homogeneity means that the "intensity" or "rate" parameter λ (i.e. number of events per unit time) is constant. Note, in particular, that $N(t)-N(s)$ is independent of $N(s)-N(0)=N(s)$. The Poisson process describes the flow of rare events. It is applied to a wide range of phenomena such as the arrival of customers in a bank agency, floods, earthquakes, car crashes, the number of particles emitted by a radioactive source, and so on (see Fig. 1.1).

From the previous assumptions, we can extract several probabilistic information.

Theorem 1.1. *The state probabilities* $p_k(t) \equiv Pr\{N(t)=k\}$, $k \geq 0$, $t \geq 0$, *satisfy the following difference-differential equations:*

$$\frac{dp_k(t)}{dt} = -\lambda p_k(t) + \lambda p_{k-1}(t), \qquad k \geq 0, t > 0 \tag{1.1}$$

with initial conditions

$$p_k(0) = \begin{cases} 1 & k=0 \\ 0 & k \geq 1 \end{cases} \qquad \textit{and} \quad p_{-1}(t) = 0 \quad \forall t \tag{1.2}$$

Proof. In order to derive the equations (1.1)–(1.2), we write

$$\begin{aligned} p_k(t+dt) &= \\ &= \Pr\{N(t+dt] = k\} = \\ &= \Pr\Big\{(N(t)=k, N(t,t+dt]=0) \bigcup \\ &\bigcup (N(t)=k-1, N(t,t+dt]=1) \bigcup \\ &\textstyle\bigcup_{j=2}^{k} (N(t)=k-j, N(t,t+dt]=j)\Big\} = \\ &= \Pr\{N(t)=k, N(t,t+dt]=0\} + \\ &+ \Pr\{N(t)=k-1, N(t,t+dt]=1\} + \\ &+ \sum_{j=2}^{k} \Pr\{N(t)=k-j, N(t,t+dt]=j\} = \\ &= p_k(t)(1-\lambda dt) + p_{k-1}(t)\lambda dt + o(dt) \end{aligned} \tag{1.3}$$

From (1.3), equation (1.1) immediately emerges. □

Theorem 1.2. *The distribution of the homogeneous Poisson process is given by*

$$p_k(t) = Pr\{N(t) = k \mid N(0) = 0\} = e^{-\lambda t}\frac{(\lambda t)^k}{k!}, \qquad k \geq 0,\ t > 0 \tag{1.4}$$

which is called Poisson distribution with parameter λt.

Proof. The simplest method to obtain (1.4) is the technique based on the *probability generating function* (for the p.g.f. see Appendix A)

$$G(u,t) = \sum_{k=0}^{\infty} u^k \cdot p_k(t), \qquad t \geq 0,\ |u| \leq 1. \tag{1.5}$$

From (1.1) we can write

$$\begin{aligned}\sum_{k=0}^{\infty} \frac{\partial p_k(t)}{\partial t} u^k &= \sum_{k=0}^{\infty}(-\lambda p_k(t) + \lambda p_{k-1}(t))u^k \\ = -\lambda G(u,t) + \lambda \sum_{k=0}^{\infty} p_{k-1}(t) u^k &= -\lambda G(u,t) + \lambda \sum_{h=0}^{\infty} p_h(t) u^{h+1} \\ &= -\lambda G(u,t) + \lambda u G(u,t) = \lambda(u-1)G(u,t)\end{aligned} \tag{1.6}$$

thus obtaining the Cauchy problem:

$$\begin{cases} \frac{\partial}{\partial t} G(u,t) = \lambda(u-1)G(u,t) \\ G(u,0) = 1. \end{cases} \tag{1.7}$$

The general solution of (1.7) is given by

$$G(u,t) = k e^{\lambda t(u-1)}, \tag{1.8}$$

and since

$$G(u,0) = 1 \tag{1.9}$$

we obtain that

$$G(u,t) = e^{\lambda t(u-1)} = e^{-\lambda t} \sum_{k=0}^{\infty} \frac{(\lambda t)^k}{k!} u^k, \qquad t > 0, |u| \leq 1. \tag{1.10}$$

Comparing (1.5) and (1.10), we extract the distribution (1.4).

□

Note that it sums to 1: $\sum_{k=0}^{\infty} e^{-\lambda t}\frac{(\lambda t)^k}{k!} = e^{-\lambda t}e^{\lambda t} = 1$, and, for any t, $N(t)$ is a Poisson r.v. with parameter λt.

Its *moment generating function* (m.g.f.) is

$$\mathbb{E}[X] = \sum_{k=0}^{\infty} e^{\theta k} \cdot p_k(t) =$$
$$= \sum_{k=0}^{\infty} e^{\theta k} e^{-\lambda t}\frac{(\lambda t)^k}{k!} = e^{-\lambda t}\sum_{k=0}^{\infty}\frac{(e^{\theta}\lambda t)^k}{k!} = \tag{1.11}$$
$$= e^{-\lambda t}e^{e^{\theta}\lambda t} = \exp\{\lambda t(e^{\theta}-1)\}$$

Remark 1.1. The Poisson distribution as limit of the binomial distribution

It is possible to obtain the Poisson distribution as a "special case" of the Binomial distribution when $n \to \infty$, $p_n \to 0$ and $np_n = \lambda$.

Indeed, being $p_n = \frac{\lambda}{n}$

$$\lim_{\substack{n\to\infty \\ p_n\to 0}} \binom{n}{k} p_n^k (1-p_n)^{n-k} =$$
$$= \lim_{n\to\infty} \tfrac{n(n-1)\dots(n-k+1)}{k!}\left(\tfrac{\lambda}{n}\right)^k\left(1-\tfrac{\lambda}{n}\right)^{n-k} =$$
$$= \lim_{n\to\infty} \tfrac{n^k+O(n^{k-1})}{k!}\left(\tfrac{\lambda}{n}\right)^k\left(1-\tfrac{\lambda}{n}\right)^{n-k} =$$
$$= \lim_{n\to\infty} \tfrac{\lambda^k}{k!}\tfrac{(1-\frac{\lambda}{n})^n}{(1-\frac{\lambda}{n})^k} = e^{-\lambda}\tfrac{\lambda^k}{k!}$$

As observed by Sheldon M. Ross (2024, p. 70), the Poisson limit holds not only if the n Bernoulli r.v.'s are i.i.d., but also if they are not identically distributed nor independent, provided that their dependence is sufficiently "weak".

Theorem 1.3. *An alternative method for the derivation of* (1.4) *recursively is to resolve the equations* (1.1).

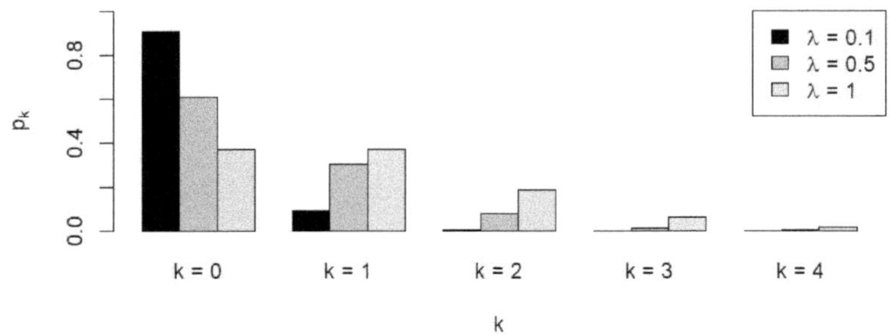

FIGURE 1.1
Probability Mass Function of the Poisson distribution, for $\lambda \in \{0.1, 0.5, 1\}$.

Proof. The solution of

$$\begin{cases} \frac{dp_0(t)}{dt} = -\lambda p_0(t), \\ p_0(0) = 1, \end{cases} \tag{1.12}$$

is $p_0(t) = e^{-\lambda t}$ with $t > 0$. If we suppose that

$$p_{k-1}(t) = e^{-\lambda t}\frac{(\lambda t)^{k-1}}{(k-1)!}, \quad k \geq 1 \tag{1.13}$$

we obtain the following non-homogeneous linear equation

$$\frac{dp_k(t)}{dt} = -\lambda p_k(t) + \lambda e^{-\lambda t}\frac{(\lambda t)^{k-1}}{(k-1)!} \tag{1.14}$$

with the initial condition $p_k(0) = 0$ for $k > 0$. The equation (1.14) has the form

$$y'(t) + \alpha(t)y(t) = \beta(t) \tag{1.15}$$

and its general solution reads

$$y(t) = e^{-\int_0^t \alpha(u)du}\left\{\int_0^t \beta(u)e^{\int_0^z \alpha(w)dw} + \text{const}\right\}. \tag{1.16}$$

So, after some calculation we have that

$$
\begin{aligned}
p_k(t) = e^{-\lambda t}\left\{\lambda \int_0^t e^{\lambda s} e^{-\lambda s}\frac{(\lambda s)^{k-1}}{(k-1)!}ds + \text{const}\right\} = \\
= e^{-\lambda t}\left\{\frac{(\lambda t)^k}{k!} + \text{const}\right\} = e^{-\lambda t}\frac{(\lambda t)^k}{k!},
\end{aligned}
\tag{1.17}
$$

where the initial condition is $p_k(0) = 0$, $k \geq 1$, has been taken into account. □

1.1.1 Properties of the Homogeneous Poisson Process

(i) Under the condition that $N(t) = n$, the conditional distribution of the instants (epochs) of occurrence of the n events $T_1, \cdots, T_n$ is

$$
\Pr\{T_1 \in dt_1, \cdots, T_n \in dt_n \mid N(t) = n\} = \frac{n!}{t^n} dt_1 \cdots dt_n, \quad \text{with } 0 < t_1 < t_2 < \cdots < t_n < t
\tag{1.18}
$$

Proof. From property 5 in Definition 1 and formula (1.4), we have that

$$
\begin{aligned}
&Pr\{T_1 \in dt_1, \cdots, T_n \in dt_n \mid N(t) = n\} = \\
&= \frac{Pr\{T_1 \in dt_1, \cdots, T_n \in dt_n, N(t) = n\}}{Pr\{N(t) = n\}} = \\
&= \frac{1}{Pr\{N(t) = n\}} \cdot Pr\{N(0, t_1] = 0, N(t_1, t_1 + dt_1] = 1, \cdots, \\
&N(t_{n-1}, t_n] = 0, N(t_n, t_n + dt_n] = 1, N(t_n + dt_n, t] = 0\} = \\
&= \frac{\overbrace{e^{-\lambda t_1}}^{\text{no jumps}} \overbrace{\lambda dt_1}^{\text{jump}} \cdots \overbrace{e^{-\lambda(t_n - t_{n-1})}}^{\text{no jumps}} \overbrace{\lambda dt_n}^{\text{jump}} \overbrace{e^{-\lambda(t - t_n)}}^{\text{no jumps}}}{\frac{(\lambda t)^n}{n!} e^{-\lambda t}} = \\
&= n! \frac{dt_1}{t} \dots \frac{dt_n}{t} = \frac{n!}{t^n} dt_1 \cdots dt_n \quad \square
\end{aligned}
\tag{1.19}
$$

and thus $(T_1, ..., T_n)$ is uniformly distributed over a simplex.

In other words, given $N(t) = n$, the n arrival times $T_1, ..., T_n$ have a joint conditional distribution $\frac{n!}{t^n}$ which is the distribution of the order statistics of n independent r.v.'s uniformly distributed on $[0, t]$.
Note that if $T_n = t$ the result holds for $T_1, ..., T_{n-1}$.

The following scheme helps explaining the intermediate step of (1.19):

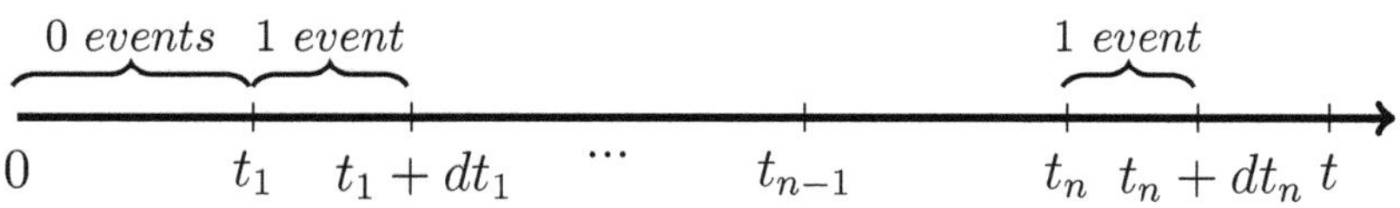

Note that if $N(t) = 1$, we have that

$$\Pr\{T_1 \in dt_1 \mid N(t) = 1\} = \frac{\Pr\{T_1 \in dt_1, N(t) = 1\}}{\Pr\{N(t) = 1\}} = \\ = \frac{e^{-\lambda t_1}\lambda dt_1 e^{-\lambda(t-t_1)}}{\lambda t e^{-\lambda t}} = \frac{dt_1}{t} \tag{1.20}$$

and thus T_1 is uniformly distributed in $(0, t)$.
Note that $\Pr\{T_1 > t\} = \Pr\{N(t) = 0\} = p_0(t) = e^{-\lambda t}$ and it is called the "survival probability".

The joint *unconditional* probability is the numerator of (1.19)

$$\Pr\{T_1 \in dt_1, ..., T_n \in dt_n\} = \lambda^n e^{-\lambda t} dt_1 ... dt_n \tag{1.21}$$

Taking logarithms and maximizing, we find the maximum likelihood estimator (MLE, see § 1.1.5.2) of λ as

$$\lambda_{MLE} = \frac{n}{t} \tag{1.22}$$

Example (Ross, 2024, p. 342). Insurance claims are represented by means of a homogeneous Poisson process of rate $\lambda > 0$. The claim amounts are independent i.d.

random variables C_i, $i \geq 1$, with distribution function F_i. By time T_i we denote the instant of occurrences of the i-th claim.

The total discounted value of all claims up to time t as

$$D(t) = \sum_{i=1}^{N(t)} e^{-rT_i} C_i \tag{1.23}$$

where r is the discount rate.

Then

$$\mathbb{E}[D(t)] = \sum_{n=0}^{\infty} \mathbb{E}[D(t) \mid N(t) = n] e^{-\lambda t} \frac{(\lambda t)^n}{n!} \tag{1.24}$$

and

$$\begin{aligned} \mathbb{E}[D(t) \mid N(t) = n] &= \mathbb{E}\left[\sum_{i=1}^{n} e^{-rT_i} C_i\right] = \\ &= \sum_{i=1}^{n} \mathbb{E}[e^{-rT_i} C_i] = \sum_{i=1}^{n} \mathbb{E}[C_i]\mathbb{E}[e^{-rT_i}] \end{aligned} \tag{1.25}$$

assuming $\mathbb{E}[C_i] = \mu$ we have

$$(\star) = \mu \sum_{i=1}^{n} \mathbb{E}[e^{-rT_i}] = \mu \mathbb{E}\left[\sum_{i=1}^{n} e^{-rT_i}\right] = \mu \mathbb{E}\left[\sum_{i=1}^{n} e^{-rU_i}\right] \tag{1.26}$$

where U_i are Uniform r.v.'s in $[0, t]$

$$(\star) = n\mu \mathbb{E}[e^{-rU}] = n\mu \int_0^t e^{-ru} \frac{1}{t} du = \frac{n\mu}{rt}(1 - e^{-rt}) \tag{1.27}$$

Finally:

$$\mathbb{E}[D(t)] = \sum_{n=0}^{\infty} \frac{n\mu}{rt}(1-e^{-rt})e^{-\lambda t}\frac{(\lambda t)^n}{n!} = \frac{\lambda\mu}{r}(1-e^{-rt}) \tag{1.28}$$

(ii) The mean and variance of the Poisson process are[1]

$$\begin{cases} \mathbb{E}[N(t)] = \lambda t \\ \mathbb{V}ar[N(t)] = \lambda t \end{cases} \tag{1.29}$$

In fact

$$\begin{aligned} \mathbb{E}[N(t)] &= \sum_{h=0}^{\infty} h e^{-\lambda t}\frac{(\lambda t)^h}{h!} = \\ &= \sum_{h=1}^{\infty} h e^{-\lambda t}\frac{(\lambda t)^h}{h!} = e^{-\lambda t}\sum_{h=1}^{\infty}\frac{(\lambda t)^h}{(h-1)!} = \\ &= e^{-\lambda t}\lambda t\sum_{h=1}^{\infty}\frac{(\lambda t)^{h-1}}{(h-1)!} = e^{-\lambda t}\lambda t\sum_{m=0}^{\infty}\frac{(\lambda t)^m}{m!} = e^{-\lambda t}\lambda t e^{\lambda t} = \lambda t \end{aligned} \tag{1.30}$$

and

$$\begin{aligned} \mathbb{E}[N(t)^2] &= \sum_{h=0}^{\infty} h^2 e^{-\lambda t}\frac{(\lambda t)^h}{h!} = e^{-\lambda t}\sum_{h=1}^{\infty} h\frac{(\lambda t)^h}{(h-1)!} = \\ &= e^{-\lambda t}\lambda t\sum_{m=0}^{\infty}(m+1)\frac{(\lambda t)^m}{m!} = \\ &= e^{-\lambda t}\lambda t\left[\sum_{m=0}^{\infty} m\frac{(\lambda t)^m}{m!} + e^{\lambda t}\right] = \\ &= e^{-\lambda t}\lambda t\left[e^{\lambda t}\sum_{m=0}^{\infty} m e^{-\lambda t}\frac{(\lambda t)^m}{m!} + e^{\lambda t}\right] = \\ &= e^{-\lambda t}\lambda t\left[e^{\lambda t}\lambda t + e^{\lambda t}\right] = \lambda t(\lambda t+1) \end{aligned} \tag{1.31}$$

[1]See § 1.1.3 about *overdispersion*.

hence

$$\mathbb{V}ar[N(t)] = (\lambda t)^2 + \lambda t - (\lambda t)^2 = \lambda t \tag{1.32}$$

Analogously, it can be proved that the third moment about the mean (third central moment) is

$$\mu_3^c \equiv \mathbb{E}\left[(N(t) - \lambda t)^3\right] = \lambda t \tag{1.33}$$

and the fourth central moment

$$\mu_4^c \equiv \mathbb{E}\left[(N(t) - \lambda t)^4\right] = 3(\lambda t)^2 + \lambda t \tag{1.34}$$

In fact, the third central moment

$$\begin{aligned} \mathbb{E}\left[(N(t) - \lambda t)^3\right] &= \\ = \mathbb{E}\left[N^3(t) - 3\lambda t N^2(t) + 3(\lambda t)^2 N(t) - (\lambda t)^3\right] &= \\ = \mathbb{E}\left[N^3(t)\right] - 3\lambda t\left[\lambda t(\lambda t + 1) + 3(\lambda t)^2 \lambda t - (\lambda t)^3\right] &= \\ = \mathbb{E}\left[N^3(t)\right] - 3(\lambda t)^3 - 3(\lambda t)^2 + 3(\lambda t)^3 - (\lambda t)^3 &= \\ = \mathbb{E}\left[N^3(t)\right] - (\lambda t)^3 - 3(\lambda t)^2 &= \\ &= \lambda t \end{aligned} \tag{1.35}$$

where $\mathbb{E}\left[N^3(t)\right] = (\lambda t)^3 + 3(\lambda t)^2 + \lambda t$.

In fact, since we can write $\frac{d}{du}\left(\mathbb{E}\left[u^{N(t)}\right]\right) = \mathbb{E}\left[N(t)u^{N(t)-1}\right]$, that for $u = 1$ equals $\mathbb{E}\left[N(t)\right] = \lambda t$, it also turns out that

$$\begin{aligned} \frac{d^3}{du^3}(\mathbb{E}[u^{N(t)}])_{|u=1} &= \\ = \mathbb{E}[N(t)(N(t) - 1)(N(t) - 2)] &= \\ = \mathbb{E}[N^3(t) - 3N^2(t) + 2N(t)] &= \\ = \mathbb{E}[N^3(t) - 3\lambda t(\lambda t + 1) + 2\lambda t] &= \\ = \mathbb{E}[N^3(t)] - 3(\lambda t)^2 - \lambda t &= \\ &= (\lambda t)^3 \end{aligned} \tag{1.36}$$

Therefore, the skewness is

$$sk\,(N(t)) = \frac{\mu_3^c}{(\mu_2^c)^{\frac{3}{2}}} = \frac{\lambda t}{(\lambda t)^{\frac{3}{2}}} = \frac{1}{\sqrt{\lambda t}} \tag{1.37}$$

and the kurtosis excess is

$$ku(N(t)) = \frac{\mu_4^c}{(\mu_2^c)^2} - 3 = \frac{3(\lambda t)^2 + \lambda t}{(\lambda t)^2} - 3 = \frac{1}{\lambda t} \tag{1.38}$$

Remark 1.2.
Note that $N(t) \in \mathbb{N}$ so that

$$\begin{aligned}\mathbb{E}[N(t)] &= \sum_{k=1}^{\infty} k \cdot \Pr\{N(t) = k\} = \\ &= \sum_{k=1}^{\infty}\sum_{j=1}^{k} \Pr\{N(t) = k\} =\end{aligned}$$

being $(k \geq 1,\, 1 \leq j \leq k)$ equivalent to $(j \geq 1,\, k \geq j)$ (1.39)

$$\begin{aligned}&= \sum_{j=1}^{\infty}\sum_{k=j}^{\infty} \Pr\{N(t) = k\} = \sum_{j=1}^{\infty} \Pr\{N(t) \geq j\} = \\ &= \sum_{j=1}^{\infty} \bar{F}_{N(t)}(j)\end{aligned}$$

The event $\{N(t) \geq j\}$ is equivalent to $\{T_j \leq t\}$. Therefore

$$\mathbb{E}[N(t)] = \sum_{j=1}^{\infty} \Pr\{T_j \leq t\} = \sum_{j=1}^{\infty} F_{T_j}(t) \tag{1.40}$$

T_j being the epoch of the j-th arrival.

(iii) The covariance is for $s < t$

$$\begin{aligned} \mathbb{C}ov[N(s), N(t)] &= \mathbb{E}[N(s)N(t)] - \mathbb{E}[N(s)]\mathbb{E}[N(t)] = \\ &= \mathbb{E}[N(s)(N(t) - N(s) + N(s))] - \mathbb{E}[N(s)\mathbb{E}[N(t)]] = \\ &= \mathbb{E}[N^2(s)] + \mathbb{E}[N(s)(N(t) - N(s))] - \mathbb{E}[N(s)]\mathbb{E}[N(t)] = \\ &= (\lambda s + \lambda^2 s^2) + (\lambda s \lambda (t-s)) - (\lambda s \lambda t) = \\ &= \lambda s = \lambda(s \wedge t) \end{aligned} \tag{1.41}$$

(iv) The first-passage time at level k is

$$T_k = \inf\{s : N(s) = k\} \tag{1.42}$$

with distribution

$$\begin{aligned} \Pr\{T_k \in ds\} &= \Pr\{N(s) = k-1, N(s, s+ds] = 1\} = \\ &= e^{-\lambda s}\frac{(\lambda s)^{k-1}}{(k-1)!}\lambda ds = \frac{\lambda^k}{\Gamma(k)}e^{-\lambda s}s^{k-1}ds \end{aligned} \tag{1.43}$$

and

$$\begin{aligned} \Pr\{T_k < \infty\} &= \\ &= \int_0^\infty e^{-\lambda s}\frac{(\lambda s)^{k-1}}{(k-1)!}\lambda ds = \\ &= \frac{\lambda^k}{(k-1)!}\int_0^\infty s^{k-1}e^{-\lambda s}ds = \\ &= \frac{\lambda^k}{(k-1)!}\int_0^\infty \left(\frac{w}{\lambda}\right)^{k-1} e^{-w}\frac{dw}{\lambda} = \\ &= \frac{1}{(k-1)!}\int_0^\infty w^{k-1}e^{-w}dw = \\ &= \frac{1}{(k-1)!}\Gamma(k) = 1. \end{aligned} \tag{1.44}$$

This means that the Poisson process reaches any level k in a finite time with probability one.

(v)

Theorem 1.4. *The Poisson process $N(t)$ has stationary increments in the sense that the distribution of the increment $N(t) - N(s)$, $t > s > 0$ depends only on $(t-s)$*

$$Pr\{N(t) - N(s) = k\} = e^{-\lambda(t-s)} \frac{\lambda^k (t-s)^k}{k!} \tag{1.45}$$

Therefore, $N(t) - N(s)$ has the distribution of $N(t-s)$.

Proof.

$$\begin{aligned}
\Pr\{N(t) = k\} &= e^{-\lambda t} \frac{(\lambda t)^k}{k!} = \\
&= \Pr\{N(t) - N(s) + N(s) = k\} = \\
&= \sum_{h=0}^{k} \Pr\{N(t) - N(s) = k-h, N(s) = h\} = \\
&\quad \text{(by independence of the increments)} \\
&= \sum_{h=0}^{k} \Pr\{N(t) - N(s) = k-h\} \Pr\{N(s) = h\} = \\
&= \sum_{h=0}^{k} e^{-\lambda s} \frac{(\lambda s)^h}{h!} \Pr\{N(t) - N(s) = k-h\}
\end{aligned} \tag{1.46}$$

and the unique solution of the previous equation is (1.45). In fact

$$\begin{aligned}
&\sum_{h=0}^{k} e^{-\lambda s} \frac{(\lambda s)^h}{h!} e^{-\lambda(t-s)} \frac{\lambda^{k-h} (t-s)^{k-h}}{(k-h)!} = \\
&= e^{-\lambda t} \frac{\lambda^k}{k!} \sum_{h=0}^{k} k! \frac{s^h (t-s)^{k-h}}{h!(k-h)!} = \\
&\quad \text{(by the binomial theorem)} \\
&= e^{-\lambda t} \frac{\lambda^k}{k!} [s + (t-s)]^k = e^{-\lambda t} \frac{(\lambda t)^k}{k!}
\end{aligned} \tag{1.47}$$

□

This property characterizes the Poisson process in the sense that the homogeneous Poisson process can be defined as a continuous-time, discrete-space process with independent increments such that

$$N(0) = 0 \qquad a.s. \tag{1.48}$$

and (1.45) holds.

Remark 1.3.
Note that sampling a Poisson process at fixed time intervals of length h implies, $\forall h$

$$\mathbb{E}[N(t) - N(t-h)] = \mathbb{V}ar[N(t) - N(t-h)] = h\lambda. \tag{1.49}$$

Therefore, to test the Poisson hypothesis:

$$\mathbb{E}\left[\frac{N(t) - N(t-h)}{h}\right] = \mathbb{V}ar\left[\frac{N(t) - N(t-h)}{\sqrt{h}}\right] = \lambda, \forall h \tag{1.50}$$

(vi) The following property relates the Poisson *process* with the Binomial distribution:

Theorem 1.5. *For any* $0 < s < t$ *and given* $N(0,t] = k$*, it holds true that*

$$Pr\{N(0,s] = h \mid N(0,t] = k\} = \binom{k}{h}\left(\frac{s}{t}\right)^h\left(1 - \frac{s}{t}\right)^{k-h}, \quad 0 \leq h \leq k \tag{1.51}$$

which is a Binomial distribution with parameters $(k, \frac{s}{t})$.

Proof. By taking into account the independence and the homogeneity of the increments of $N(t)$ (property 5), we can write

$$\begin{aligned}
&\Pr\{N(0,s] = h \mid N(0,t] = k\} = \\
&= \frac{\Pr\{N(0,s] = h, N(0,t] = k\}}{\Pr\{N(0,t] = k\}} = \\
&= \frac{\Pr\{N(0,s] = h, N(s,t] = k-h\}}{\Pr\{N(0,t] = k\}} = \\
&= \frac{\Pr\{N(0,s] = h\}\Pr\{N(s,t] = k-h\}}{\Pr\{N(0,t] = k\}} = \\
&= \frac{e^{-\lambda s}\frac{(\lambda s)^h}{h!} e^{-\lambda(t-s)}\frac{[\lambda(t-s)]^{k-h}}{(k-h)!}}{e^{-\lambda t}\frac{(\lambda t)^k}{k!}}
\end{aligned} \tag{1.52}$$

From (1.52), result (1.51) immediately follows. □

(vii) Let $N_1(t) \sim Poi(\lambda_1)$ and $N_2(t) \sim Poi(\lambda_2)$ independent Poisson processes. Then the sum $Y(t) = N_1(t) + N_2(t) \sim Poi(\lambda_1 + \lambda_2)$.

In fact $\{N_1 + N_2 = n\} = \cup_{k=0}^{n}\{N_1 = k, N_2 = n-k\}$, i.e. the union of disjoint events, so that

$$\begin{aligned}
\Pr\{N_1 + N_2 = n\} &= \sum_{k=0}^{n} \Pr\{N_1 = k, N_2 = n-k\} = \\
&= \sum_{k=0}^{n} \Pr\{N_1 = k\}\Pr\{N_2 = n-k\} = \\
&= \sum_{k=0}^{n} e^{-\lambda_1 t}\frac{(\lambda_1 t)^k}{k!} e^{-\lambda_2 t}\frac{(\lambda_2 t)^{n-k}}{(n-k)!} = \\
&= \frac{e^{-(\lambda_1+\lambda_2)t}}{n!}\sum_{k=0}^{n} \frac{n!}{k!(n-k)!}(\lambda_1 t)^k(\lambda_2 t)^{n-k} = \\
&\qquad \text{(by the binomial theorem)} \\
&= \frac{e^{-(\lambda_1+\lambda_2)t}}{n!}(\lambda_1 t + \lambda_2 t)^n
\end{aligned} \tag{1.53}$$

In general, the sum of independent Poisson processes $N_i(t)$ with rates λ_i, $i = 1, ..., n$ is a Poisson r.v. with rate $\Lambda = \sum_{i=1}^{n} \lambda_i$. Indeed, the probability generating function of $Y = \sum_{i=1}^{n} N_i(t)$ is

$$G_Y(u) = \mathbb{E}\left[u^Y\right] = \mathbb{E}\left[u^{\sum_{i=1}^{n} N_i(t)}\right] = \prod_{i=1}^{n} \mathbb{E}\left[u^{N_i(t)}\right] =$$
$$= \prod_{i=1}^{n} e^{-\lambda_i(1-u)} = e^{-\sum_{i=1}^{n} \lambda_i(1-u)} = e^{-\Lambda(1-u)} \tag{1.54}$$

and thus Y is a Poisson r.v. with rate $\Lambda = \sum_{i=1}^{n} \lambda_i$.

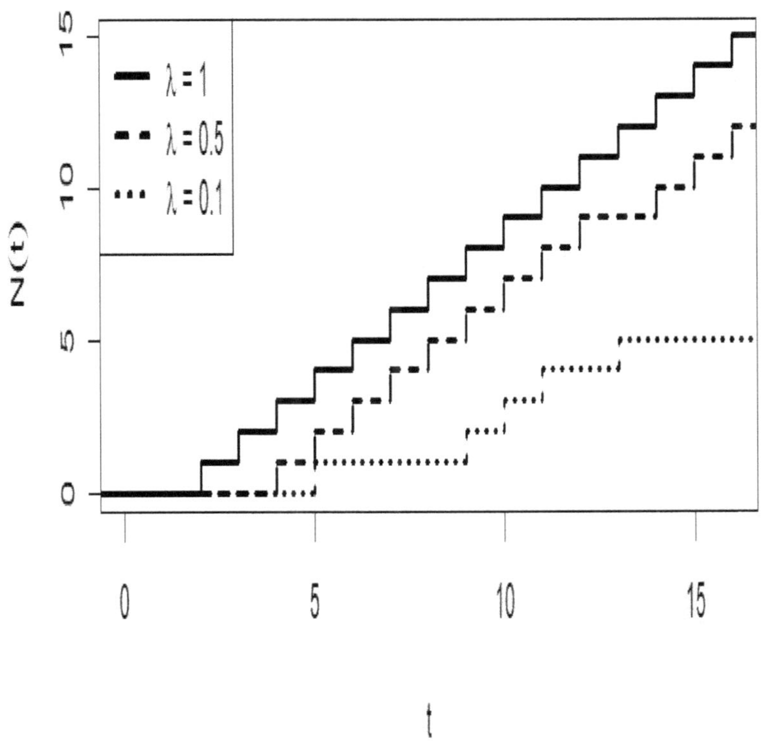

FIGURE 1.2
Some homogeneous Poisson paths for $\lambda = 1$, $\lambda = 0.5$ and $\lambda = 0.1$.

Exercise (Ross, 2024, pp. 104-106). Calculate the conditional probability of $N_1(t) \mid N_1(t) + N_2(t) = n$ where N_1, N_2 are independent Poisson processes, $N_1 \sim Poi(\lambda_1)$ and $N_2 \sim Poi(\lambda_2)$.

Solution. The given conditional probability reads

$$
\begin{aligned}
\Pr\{N_1 = k \mid N_1 + N_2 = n\} = \frac{\Pr\{N_1 = k, N_1 + N_2 = n\}}{\Pr\{N_1 + N_2 = n\}} = \\
= \frac{\Pr\{N_1 = k, N_2 = n-k\}}{\Pr\{N_1 + N_2 = n\}} = \frac{\Pr\{N_1 = k\}\Pr\{N_2 = n-k\}}{\Pr\{N_1 + N_2 = n\}} = \\
= e^{-\lambda_1 t}\frac{(\lambda_1 t)^k}{k!} e^{-\lambda_2 t}\frac{(\lambda_2 t)^{n-k}}{(n-k)!}\left[e^{-(\lambda_1+\lambda_2)t}\frac{((\lambda_1+\lambda_2)t)^n}{n!}\right]^{-1} = \\
= \frac{n!}{k!(n-k)!}\frac{(\lambda_1 t)^k(\lambda_2 t)^{n-k}}{(\lambda_1 t + \lambda_2 t)^n} = \\
= \binom{n}{k}\left(\frac{\lambda_1}{\lambda_1+\lambda_2}\right)^k\left(\frac{\lambda_2}{\lambda_1+\lambda_2}\right)^{n-k}
\end{aligned}
\tag{1.55}
$$

so that it is Binomial with parameters $\left(n, \frac{\lambda_1}{\lambda_1+\lambda_2}\right)$ and $\mathbb{E}[N_1(t) \mid N_1(t) + N_2(t) = n] = n\frac{\lambda_1}{\lambda_1+\lambda_2}$.

Theorem 1.6. *If $N(t)$ is a Poisson process with intensity λ and if each event is independently classified in k types with probability p_i, $i = 1, ..., k$, $\sum_{i=1}^k p_i = 1$, then the number of each type $N_i(t)$ is an independent Poisson process with intensity λp_i.*

Proof. Let $n = \sum_{i=1}^k n_i$ and $t = 1$ for simplicity. Then

$$
\begin{aligned}
\Pr\{N_1(t) = n_1, ..., N_k(t) = n_k\} = \\
= \Pr\{N_1(t) = n_1, ..., N_k(t) = n_k \mid N(t) = n\}\Pr\{N(t) = n\} = \\
= \frac{n!}{n_1!...n_k!}p_1^{n_1}...p_k^{n_k} e^{-\lambda}\frac{\lambda^n}{n!} = \\
= \prod_{i=1}^k e^{-\lambda p_i}\frac{(\lambda p_i)^{n_i}}{n_i!}
\end{aligned}
\tag{1.56}
$$

□

Exercise (Ross, 2024, p. 149). Let $N_1 = N(t_1)$ the number of car accidents of a policyholder in year 1 and $N_2 = N(t_2) - N(t_1)$ the number of accidents in year 2. Policyholders are of k risk types with probability p_i, $\sum_{i=1}^{k} p_i = 1$ and N_1, N_2 are Poisson r.v.'s with intensity λ_i. Calculate the expected number of accidents in year 2 of the policyholder, knowing that she had n accidents in year 1.

Solution. Conditioning on the risk types

$$
\begin{aligned}
&\mathbb{E}[N_2 \mid N_1 = n] = \\
&= \sum_{j=1}^{k} \mathbb{E}[N_2 \mid R = j, N_1 = n]\Pr(R = j \mid N_1 = n) = \\
&= \sum_{j=1}^{k} \mathbb{E}[N_2 \mid R = j]\Pr(R = j \mid N_1 = n) = \\
&= \sum_{j=1}^{k} \lambda_j \frac{\Pr(R = j, N_1 = n)}{\Pr(N_1 = n)} = \\
&= \frac{\sum_{j=1}^{k} \lambda_j \Pr(N_1 = n \mid R = j)\Pr(R = j)}{\sum_{j=1}^{k} \Pr(N_1 = n \mid R = j)\Pr(R = j)} = \\
&= \frac{\sum_{j=1}^{k} \lambda_j e^{-\lambda_j} \frac{\lambda_j^n}{n!} p_j}{\sum_{j=1}^{k} e^{-\lambda_j} \frac{\lambda_j^n}{n!} p_j} = \\
&= \frac{\sum_{j=1}^{k} e^{-\lambda_j} \lambda_j^{n+1} p_j}{\sum_{j=1}^{k} e^{-\lambda_j} \lambda_j^{n} p_j}
\end{aligned}
\tag{1.57}
$$

1.1.2 Alternative Definitions of the Poisson Process

An alternative, equivalent definition of the homogeneous Poisson process is the following

Definition 2

1*. $Pr\{N(0) = 0\} = 1$.
2*. $Pr\{N(t) = k\} = e^{-\lambda t}\frac{(\lambda t)^k}{k!}$.
3*. *The process has stationary, independent increments.*

Theorem 1.7. *From Definition 2 we obtain Definition 1 and they are equivalent.*

Proof. By stationarity of increments:

$$\begin{aligned} Pr\{N(t,t+dt] = 0\} &= Pr\{N(t+dt) - N(t) = 0\} = \\ &= e^{-\lambda dt}\frac{(\lambda dt)^0}{0!} = e^{-\lambda dt} = 1 - \lambda dt + o(dt), \\ Pr\{N(t,t+dt] = 1\} &= e^{-\lambda dt}\frac{(\lambda dt)^1}{1!} = \lambda dt e^{-\lambda dt} = \\ &= \lambda dt(1 - \lambda dt + o(dt)) = \lambda dt + o(dt) \end{aligned} \tag{1.58}$$

Therefore, given Theorem 1.2, Definition 1 and Definition 2 are equivalent.

□

A third equivalent definition makes use of the concept of inter-arrival intervals. Let T_k be the time (epoch) of the k-th event ($T_0 = 0$). We define the n-th interarrival interval τ_n as

$$\tau_n \equiv T_n - T_{n-1} \tag{1.59}$$

so that $\tau_1 = T_1$ and $T_n = \sum_{k=1}^{n} \tau_k$.

Note that τ_1 is not a proper inter-arrival interval, but it has Exponential distribution being $\Pr\{\tau_1 \leq t\} = 1 - \Pr\{\tau_1 > t\} = 1 - \Pr\{N(t) = 0\} = 1 - e^{-\lambda t}$.

Definition 3.
1^{**}. $Pr\{N(0) = 0\} = 1$.
2^{**}. *The interarrival times* τ_n *are i.i.d.*
3^{**}. $Pr\{\tau_n \leq t\} = 1 - e^{-\lambda t}, \forall n \geq 1$, *i.e. they have Exponential distribution.*
4^{**}. $N(t) = \max\{n \geq 0 : T_n \leq t\}$.

Set $T_0 = 0$, $T_n = \sum_{i=1}^{n} \tau_i$ and $N(t) = \max\{n \geq 0 : T_n \leq t\}$. It can be shown that T_n is (unconditionally) $Gamma(n, \lambda)$ distributed and the three definitions are equivalent, i.e. $N(t)$ is a Poisson process.

Note the no-memory property of the Exponential distribution:

$$\begin{aligned} &Pr\{\tau > t + s \mid \tau > s\} = \\ &= \frac{Pr\{\tau > t + s, \tau > s\}}{Pr\{\tau > s\}} = \frac{Pr\{\tau > t + s\}}{Pr\{\tau > s\}} = \\ &= \frac{e^{-\lambda(t+s)}}{e^{-\lambda s}} = e^{-\lambda t} = Pr\{\tau > t\}. \end{aligned} \tag{1.60}$$

It can be written also as:

$$\Pr\{\tau > t + s\} = \Pr\{\tau > s\} \cdot \Pr\{\tau > t\} \tag{1.61}$$

This property means that, having waited up to s, the distribution of the overall waiting time $s+t$ is the same as the distribution of the remaining waiting time t: having waited up to s, it does not change the probability of future arrivals which therefore have no memory of previous waiting.

The next remark introduces *Renewal Processes* (see § 5.4).

Remark 1.4. Renewal process
Assuming for inter-arrival times a distribution different from the Exponential, we obtain a generalized process called Renewal process. In fact a Renewal Process is an arrival (point) process in which the interarrival intervals are i.i.d.

The term "renewal" process is due to the fact that the process restarts (in probabilistic terms) after every *given* arrival time T, i.e.

$$N(t+T) - N(T), \qquad t \geq 0 \tag{1.62}$$

is again a Renewal process with i.i.d. interarrival intervals and with the same distribution as $N(t)$.

If the interarrival distribution F has mean μ and variance σ^2 then, for t large (see Gallager, 1996)

$$\frac{N(t) - \frac{t}{\mu}}{\sqrt{\frac{t\sigma^2}{\mu^3}}} \sim_{t\to\infty} \Phi(0,1) \qquad \text{standard Normal} \tag{1.63}$$

Theorem 1.8. *Definition 2 and Definition 3 are equivalent.*

Proof. From *Definition 2*: if T is an arrival time, $N(t+T)-N(T) \equiv M_T(t)$ is a Poisson process for any T and by Theorem 1.4

$$\Pr\{N(t+T) - N(T) = k\} = e^{-\lambda t}\frac{(\lambda t)^k}{k!} \tag{1.64}$$

Moreover, $\tau_1 = T_1 - T_0 = T_1$ is Exponential, being

$$\begin{cases} \Pr\{\tau_1 > t\} = \Pr\{N(t) = 0\} = e^{-\lambda t} \\ \Pr\{\tau_1 \leq t\} = \Pr\{N(t) \geq 1\} = 1 - e^{-\lambda t} \end{cases} \tag{1.65}$$

In general $\tau_n = T_n - T_{n-1}$ and $N(t+T_{n-1}) - N(T_{n-1}) = M_{n-1}(t)$ is Poisson, such that

$$\begin{cases} \Pr\{\tau_n > t\} = \Pr\{M_{n-1}(t) = 0\} = e^{-\lambda t} \\ \Pr\{\tau_n \leq t\} = \Pr\{M_{n-1}(t) \geq 1\} = 1 - e^{-\lambda t} \end{cases} \tag{1.66}$$

so that the interarrival times are i.i.d. with Exponential distribution as in *Definition 3.*

Hence, from *Definition 3* and the renewal structure of the Poisson process in terms of its interarrival times, i.e.

$$\Pr\{N(t) = k\} = \Pr\{\tau_1 + ... + \tau_k < t, \tau_1 + ... + \tau_{k+1} > t\} \tag{1.67}$$

we can write[2]

$$\begin{aligned}
\Pr\{\tau_1 + ... + \tau_k < t, \tau_1 + ... + \tau_{k+1} > t\} &= \\
= \Pr\{\tau_1 + ... + \tau_{k+1} > t\} &+ \\
-\Pr\{\tau_1 + ... + \tau_{k+1} > t, \tau_1 + ... + \tau_k > t\} &= \\
= \int_t^\infty \Pr\{\tau_1 + ... + \tau_{k+1} \in ds\} - \int_t^\infty \Pr\{\tau_1 + ... + \tau_k \in ds\} &= \\
= \frac{\lambda^{k+1}}{k!} \int_t^\infty e^{-\lambda s} s^k ds - \frac{\lambda^k}{(k-1)!} \int_t^\infty e^{-\lambda s} s^{k-1} ds &
\end{aligned} \tag{1.68}$$

This yields

$$\begin{aligned}
\frac{\lambda^{k+1}}{k!} \int_t^\infty e^{-\lambda s} s^k ds - \frac{\lambda^k}{(k-1)!} \int_t^\infty e^{\lambda s} s^{k-1} ds &= \\
= \frac{\lambda^{k+1}}{k!} \int_{\lambda t}^\infty e^{-w} \left(\frac{w}{\lambda}\right)^k \frac{dw}{\lambda} - \frac{\lambda^k}{(k-1)!} \int_{\lambda t}^\infty e^{-w} \left(\frac{w}{\lambda}\right)^{k-1} \frac{dw}{\lambda} &= \\
= \frac{1}{k!} \int_{\lambda t}^\infty e^{-w} w^k dw - \frac{1}{(k-1)!} \int_{\lambda t}^\infty e^{-w} w^{k-1} dw &= \\
= \frac{1}{k!} \left\{ \left[-e^{-w} w^k\right]_{\lambda t}^\infty + \int_{\lambda t}^\infty k w^{k-1} e^{-w} dw \right\} &+ \\
- \frac{1}{(k-1)!} \int_{\lambda t}^\infty e^{-w} w^{k-1} dw &= \\
= e^{-\lambda t} \frac{(\lambda t)^k}{k!} = \Pr\{N(t) = k\} &
\end{aligned} \tag{1.69}$$

□

[2]Since the intersection of two events $\Pr\{A \cap B\}$ equals $\Pr\{A\} - \Pr\{A \cap B^c\}$.

Remark 1.5. Generating a Poisson process
The inter-arrival intervals τ_n have exponential distribution

$$p = F_{\tau_n}(t) = \Pr\{\tau_n \leq t\} = 1 - e^{-\lambda t}, \qquad \forall n \geq 1 \tag{1.70}$$

The inverse distribution is, for $0 < p < 1$

$$t = F^{-1}(p) = -\frac{\log(1-p)}{\lambda} \tag{1.71}$$

Drawing a sample of n data points from the Uniform distribution $U(0,1)$, $p_1, ..., p_n$, one obtains a sample of n interarrival intervals $\tau_1, ..., \tau_n$ so that the epoch of the j-th arrival is

$$T_j = \sum_{h=1}^{j} \tau_h, \qquad j = 1, ..., n \tag{1.72}$$

$$N(t) = \begin{cases} 0 \text{ if } t < T_1 \\ j \text{ if } T_j \leq t < T_{j+1} \\ n \text{ if } T_n \leq t \end{cases} \tag{1.73}$$

Note that τ_1 is not precisely an interarrival interval but it has Exponential distribution.

Theorem 1.9. *Let $N(t)$ be a Poisson process with intensity λ and classify its events in two subgroups with probabilities p_1 and $p_2 = 1 - p_1$. The resulting processes $N_1(t)$ and $N_2(t)$ are independent Poisson r.v.'s with intensity λp_1 and λp_2 respectively.*

Proof. See Ross (2004), p. 328 and Theor. 1.6.

1.1.3 Overdispersion

In the Poisson process mean and variance are equal. Often, however, empirical count data exhibit significant *overdispersion* (possibly, *underdispersion*), i.e. an index of dispersion $D = \frac{\sigma^2}{\mu} > 1$ ($D < 1$). In the case of overdispersion, the fit of a parametric model to data is improved by the adoption of the so called *overdispersed* (*underdispersed*) distributions, i.e. distributions such that the variance exceeds (is lower than) the mean. Such distributions can arise as the result of a *mixing distribution*, i.e. by assuming a random parameter (in the case of the Poisson distribution, a random mean): one such largely used mixture-model is the Poisson-Gamma model: the mean parameter λ is assumed to be Gamma distributed and the resulting compound distribution is the Negative Binomial, which is *overdispersed*, i.e.

Theorem 1.10. *We prove that the Negative Binomial distribution $N^{\nu,p}$ of parameters ν and p obtains as a Poisson distribution with random intensity $\lambda \sim \Gamma(\nu, \tau)$, where $p = \frac{\tau}{1+\tau}$, i.e.*

$$p_k^{NB} = \binom{k+\nu-1}{k} p^\nu (1-p)^k, \qquad k = 0, 1, 2, ...; \nu > 0 \quad (1.74)$$

where

$$g_\Gamma(\lambda) = \frac{\tau^\nu}{\Gamma(\nu)} e^{-\tau\lambda} \lambda^{\nu-1}, \qquad \tau, \nu > 0 \quad (1.75)$$

and

$$p_k^{NB}(\lambda) = \int_0^\infty \frac{e^{-\lambda}\lambda^k}{k!} \cdot g_\Gamma(\lambda) d\lambda \quad (1.76)$$

Proof. In fact, (1.74) follows after integrating (1.76):

$$\int_0^\infty \frac{e^{-\lambda}\lambda^k}{k!} \cdot \frac{\tau^\nu}{\Gamma(\nu)} e^{-\tau\lambda} \lambda^{\nu-1} d\lambda =$$

$$= \text{using the change of variable } \lambda' = \lambda(1+\tau)$$

$$= \frac{\tau^\nu}{k!\Gamma(\nu)(1+\tau)^{k+\nu}} \underbrace{\int_0^\infty e^{-\lambda'} \lambda'^{k+\nu-1} d\lambda'}_{=\Gamma(k+\nu)} = \quad (1.77)$$

$$= \frac{\Gamma(k+\nu)}{\Gamma(k+1)\Gamma(\nu)} \left(\frac{\tau}{1+\tau}\right)^\nu \left(\frac{1}{1+\tau}\right)^k$$

For $\nu = n \in \mathbb{N}$,

$$p_k^{NB} = \binom{k+n-1}{k} p^n (1-p)^k \quad (1.78)$$

□

The mean and the variance are, respectively, $\mu^{NB} = \nu\frac{1+\tau}{\tau}$ and $\sigma^2_{NB} = \frac{\nu}{\tau}\left(1+\frac{1}{\tau}\right) > \mu^{NB}$ $(\tau < 1)$.

In fact, it is well known that the first moment of a Negative Binomial distribution of parameters (p, n) obtains as n times the first moment of a Geometric distribution, i.e.

$$\begin{aligned}\mathbb{E}\left[X_G\right] = \sum_{k=1}^{\infty} kpq^{k-1} = \\ = p\sum_{k=1}^{\infty} \frac{d}{dq}q^k = p\frac{d}{dq}\sum_{k=1}^{\infty} q^k = \\ = p\frac{d}{dq}\left(\frac{1}{1-q} - 1\right) = p\frac{1}{(1-q)^2} = \frac{p}{p^2} = \frac{1}{p}\end{aligned} \tag{1.79}$$

The second moment obtains along the same line, i.e.

$$\begin{aligned}\mathbb{E}\left[X_G^2\right] = \sum_{k=1}^{\infty} k^2pq^{k-1} = \\ = p\sum_{k=1}^{\infty} k\frac{d}{dq}q^k = p\frac{d}{dq}q\cdot\sum_{k=1}^{\infty} kq^{k-1} = \\ = p\frac{d}{dq}q\cdot\sum_{k=1}^{\infty} \frac{d}{dq}q^k = p\frac{d}{dq}q\cdot\frac{d}{dq}\sum_{k=1}^{\infty} q^k = \\ = p\frac{d}{dq}q\cdot\frac{d}{dq}\left(\frac{1}{1-q} - 1\right) = p\frac{d}{dq}\frac{q}{(1-q)^2} = \\ = p\frac{1+q}{(1-q)^3} = \frac{1+q}{(1-q)^2} = \frac{1+q}{p^2}\end{aligned} \tag{1.80}$$

Then, the variance (of a Geometric r.v.) obtains as the difference between the second moment and the first moment squared, i.e.

$$\sigma_G^2 = \frac{1+q}{p^2} - \frac{1}{p^2} = \frac{q}{p^2} \tag{1.81}$$

Finally, the mean and variance of the Negative Binomial read, respectively, as $\frac{\nu}{p}$ and $\frac{\nu(1-p)}{p^2}$; w.r.t. the (ν, τ) parameters, they equal the above expressions, respectively, $\frac{\nu(1+\tau)}{\tau}$ and $\frac{\nu(1+\tau)}{\tau^2}$.

The recursive relationship holds:

$$\begin{cases} p_0 = \left(\frac{\tau}{1+\tau}\right)^{\nu} & (k=0) \\ p_{k+1} = \frac{k+\nu}{(k+1)(1+\tau)} \cdot p_k & k \geq 0 \end{cases} \tag{1.82}$$

Moreover, as ν and τ tend to ∞ (keeping μ constant) the Negative Binomial tends to the Poisson distribution. Therefore, a small value of ν is related to *mixing* or *contagion* (see § 1.1.4).

Another overdispersed distribution arising from a mixture is the Poisson-Inverse Gaussian model (PIG), where the distribution for the mean parameter λ is Inverse-Gaussian, i.e. $\lambda \sim \mathrm{IG}(\mu, h)$:

$$g_{IG}(\lambda) = \frac{\mu}{\sqrt{2\pi h \lambda^3}} e^{-\frac{(\lambda-\mu)^2}{2h\lambda}} \qquad \mu, h > 0 \tag{1.83}$$

and its probability mass function p_k is given recursively:

$$p_k^{PIG} = \begin{cases} e^{\frac{\mu}{h}\left[1-\sqrt{1+2h}\right]} & k=0 \\ \frac{\mu}{\sqrt{1+2h}} \cdot p_0 & k=1 \\ \frac{h(k-1)(2k-3)}{(1+2h)k(k-1)} \cdot p_{k-1} + \frac{\mu^2}{(1+2h)k(k-1)} \cdot p_{k-2} & k \geq 2 \end{cases} \tag{1.84}$$

The mean and the variance are, respectively, $\mu^{PIG} = \mu$ and $\sigma^2_{PIG} = \mu(1+h) > \mu^{PIG}$.

Finally, a largely used model for overdispersed data arising from the presence of excess zeros (in excess with respect to a Poisson pattern) is the Zero Inflated Poisson (ZIP) model which is, indeed, a "zero-altered" Poisson model: it arises from the combination of a (deterministic) zero-generating process (with probability α) and a Poisson process (with probability $1-\alpha$), $0 \leq \alpha < 1$, i.e.

$$p_k^{ZIP} = \alpha \cdot 1_{\{k=0\}} + (1-\alpha) \cdot \frac{e^{-\lambda}\lambda^k}{k!}, \qquad k \geq 0; \tag{1.85}$$

it can also be represented as a combination of all zero generating processes and a left truncated Poisson distribution (without zeros), i.e.

$$p_k^{ZIP} = \begin{cases} \alpha + (1-\alpha)e^{-\lambda} & k=0 \\ (1-\alpha)e^{-\lambda}\frac{\lambda^k}{k!} & k = 1, 2, 3, \ldots \end{cases} \tag{1.86}$$

An example of a distribution whose dispersion index $D < 1$ (underdispersion) is the Binomial distribution, for which $= np > np(1-p) =$ variance. Remember that the Binomial distribution tends to the Poisson distribution $Poi(\lambda)$ for $p_n = \frac{\lambda}{n} \to 0$.

1.1.4 Contagion

Consider the Poisson process $N(t)$ representing the number of accidents in $(0, t]$ for a given subject:

$$p_k(t) = e^{-\lambda t}\frac{(\lambda t)^k}{k!} \qquad k \geq 0 \tag{1.87}$$

Positive contagion means that λ is no longer constant and it increases with occurrences. For example:

$$\lambda_k = a + b \cdot k \qquad a, b > 0 \qquad (b < 0 \text{ for negative contagion}) \tag{1.88}$$

This is a case of space-non-homogeneous Poisson process.

The resulting distribution (Buhlman, 1970, p. 53) is the Negative Binomial distribution

$$p_k^{NB} = \binom{k + \alpha_k - 1}{k} p^{\alpha_n}(1-p)^k \tag{1.89}$$

where $\alpha_k = \frac{\lambda_{k+1}}{b}$ and $p = e^{-b}$.

In the terminology of Feller (1943), this is a case of "true contagion" whilst (1.74) is "apparent contagion" generated by (a continuum of) heterogeneity of the subjects.

Note that "true" and "apparent" contagion are observationally equivalent.

1.1.5 Estimation of the Poisson Process

This section contains the main issues of the Method-of-Moments and Maximum Likelihood estimators and some numerical examples follow.

1.1.5.1 Estimation Using the Method of Moments

Given a set $\underline{\mathbf{x}}$ of n observations $\{x_1, ..., x_n\}$, the method of moments equals the first m empirical moments to the first m theoretical moments, solving the resulting system of equations for the searched parameters.

In the case of the standard Poisson distribution, the estimation is trivial, as the empirical mean corresponds to the MME (see section 1.1.5.2), i.e.

$$\bar{\mu} = \bar{x} = \hat{\lambda} \tag{1.90}$$

In the case of distributions defined by two parameters, say p_1 and p_2, either the first two non-central moments or the empirical mean $\bar{x}$ and the empirical variance s^2 are set equal to their corresponding counterparts, e.g.

$$\begin{cases} \hat{\mu}_1 = \frac{1}{n}\sum_{i=1}^{n} x_i = \mu_1 \\ \hat{\mu}_2 = \frac{1}{n}\sum_{i=1}^{n} x_i^2 = \mu_2 \end{cases} \tag{1.91}$$

and

$$\begin{cases} \bar{x} = \frac{1}{n}\sum_{i=1}^{n} x_i = \mu_1 \\ s^2 = \frac{1}{n}\sum_{i=1}^{n} (x_i - \bar{x})^2 = \mu_2 - \mu_1^2 = \sigma^2 \end{cases} \tag{1.92}$$

1.1.5.2 Estimation Using Maximum Likelihood

Estimators. Given a set $\underline{\mathbf{x}}$ of n observations $\{x_1, x_2, ..., x_n\}$ drawn as i.i.d. realizations of a Poisson process of parameter λ, the corresponding likelihood function (i.e., the joint distribution of the sample) is:

$$L(\lambda; \underline{x}) = \prod_{i=1}^{n} p_x(\lambda) = \prod_{i=1}^{n} e^{-\lambda}\frac{\lambda^{x_i}}{x_i!}, \qquad 1 \leq i \leq n \tag{1.93}$$

where $p_x(\lambda) = e^{-\lambda}\frac{\lambda^{x_i}}{x_i!}$, $x_i = 0, 1, 2, ...,$ denotes the probability mass function (PMF) of the Poisson process.

It is well known that the MLE estimator is the sample average, i.e.

$$\bar{\mu} = \bar{x} = \frac{1}{n}\sum_{i=1}^{n} x_i = \hat{\lambda} \tag{1.94}$$

The *asymptotic distribution* of the MLE estimator is

$$\hat{\lambda} \sim N(\lambda, \lambda) \tag{1.95}$$

Therefore, the confidence interval at $(1-\alpha)\%$ confidence level ($\alpha = 0.05$, say) for the Poisson MLE estimator, $\hat{\lambda}$ can be constructed as

$$\bar{x} - z_{1-\frac{\alpha}{2}}\sqrt{\frac{\bar{x}}{n}} \leq \lambda \leq \bar{x} + z_{1-\frac{\alpha}{2}}\sqrt{\frac{\bar{x}}{n}} \tag{1.96}$$

as the *Fisher information* $I(\lambda) = \mathbb{E}\left[-\frac{d^2}{d\lambda^2}\ln p_x(\lambda)\right]$ for the Poisson distribution is $I(\lambda) = \frac{1}{\lambda}$, the confidence interval for a general maximum likelihood estimator $\hat{\theta}$ is related to the Fisher Information as $CI_\theta(\alpha) \equiv \left(\hat{\theta} - z_{1-\frac{\alpha}{2}}\frac{1}{\sqrt{nI(\hat{\theta})}}, \hat{\theta} + z_{1-\frac{\alpha}{2}}\frac{1}{\sqrt{nI(\hat{\theta})}}\right)$.

Estimation. As is well known, the likelihood function $L(\theta, \underline{x})$ is regarded as a function of the parameter θ, for given, fixed observations x_i, $1 \leq i \leq n$. For numeric reasons, the optimization is carried over considering the logarithm $\ln L(\theta|\underline{x})$ of the likelihood function, which is the sum of the so called *scores* $l(x_i) = \ln L\left((\theta|x_i)\right)$:

$$l(\theta|\underline{x}) = \ln\left(L(\lambda|\underline{x})\right) = \sum_{i=1}^{n} l(x_i) = -n\lambda + \sum_{i=1}^{n} x_i \ln(\lambda) - \ln(x_i!) \tag{1.97}$$

Usually the optimization is pursued by *minimizing* the *negative* of the log-likelihood function, i.e.

$$\hat{\lambda} = \text{argmin}_{\lambda>0}\{-\ln\left(L(\lambda|\underline{x})\right)\} \tag{1.98}$$

Fig. 1.3 shows the profile of the log-likelihood function, highlighting the concave shape with respect to the optimizing parameter λ.

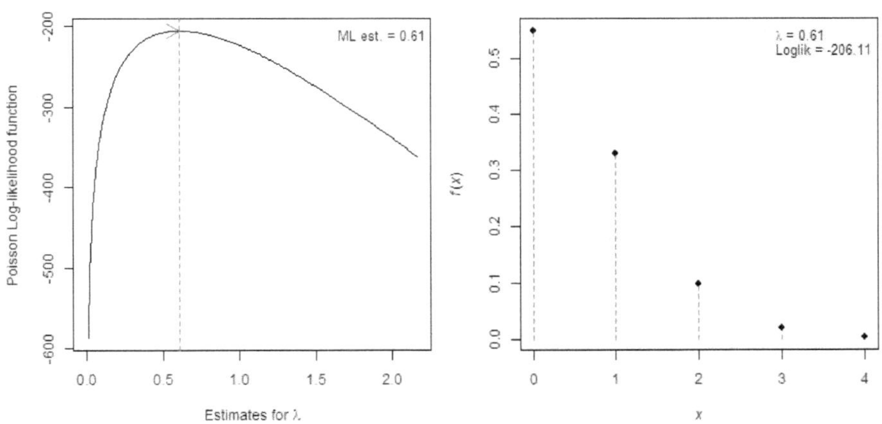

FIGURE 1.3
Maximization of the log-likelihood function (left) and corresponding PMF (right) on the data of Example I.1, $\hat{\lambda} = \bar{x} = 0.61$.

Using the interarrival times we have

$$L(\lambda; \tau_1, ..., \tau_n) = \prod_{i=1}^{n} \lambda e^{-\lambda \tau_i} = \lambda^n e^{-\lambda \sum_{i=1}^{n} \tau_i} \tag{1.99}$$

The MLE estimator is $\hat{\lambda}_{MLE} = \frac{n}{\sum_{i=1}^{n} \tau_i} = \frac{1}{\bar{\tau}}$.

Note that the mean $\bar{\tau}$ is unbiased for $\frac{1}{\lambda}$, while $\mathbb{E}[\hat{\lambda} - \lambda] = \frac{\lambda}{n-1}$ is the bias.

Remark 1.6. Testing for renewal processes
We know that a Poisson process is a point process with i.i.d. interarrival times with Exponential density $f(\tau) = \frac{1}{\mu}e^{-\frac{1}{\mu}\tau}$.

A more general case is the Gamma density (see Cox and Lewis (1966), p. 136)

$$f(\tau) = \left(\frac{k}{\mu}\right)^k \frac{1}{\Gamma(k)}\tau^{k-1}e^{-\frac{k}{\mu}\tau} \tag{1.100}$$

including the Exponential when $k = 1$.

Note that

$$\begin{cases} \mathbb{E}[\tau] = \mu \\ \mathbb{E}[\tau] = \frac{\mu^2}{k} \\ \mathbb{C}v[\tau] = \frac{\sigma}{\mu} = \frac{1}{\sqrt{k}} \\ \mathbb{S}k[\tau] = 2\mathbb{C}v[\tau] \end{cases} \tag{1.101}$$

Given n observed times $\tau_1, ..., \tau_n$ from a random sample, the likelihood function is

$$L(\underline{\tau}; k; \mu) = \left(\frac{k}{\mu}\right)^{nk} \frac{1}{\Gamma(k)^n}\prod_{i=1}^{n} \tau_i^{k-1}e^{-k\sum\limits_{i=1}^{n}\frac{\tau_i}{\mu}} \tag{1.102}$$

and the log-likelihood is

$$nk\left(\log(k) - \log(\mu)\right) + (k-1)\sum_{i=1}^{n}\log(\tau_i) - k\sum_{i=1}^{n}\frac{\tau_i}{\mu} - n\log(\Gamma(k)) \tag{1.103}$$

The ML estimators are the solution of

$$\begin{cases} -n + \sum_{i=1}^{n}\frac{\tau_i}{\mu} = 0 \\ n(\log(k) - \log(\mu)) + n + \sum\limits_{i=1}^{n}\log(\tau_i) - \sum\limits_{i=1}^{n}\frac{\tau_i}{\mu} - n\Psi(k) = 0 \end{cases} \tag{1.104}$$

Remark 1.6. (cont.)
where $\Psi(k)$ is the digamma function

$$\Psi(k) = \frac{d\log(\Gamma(k))}{dk} = \frac{\Gamma'(k)}{\Gamma(k)} \tag{1.105}$$

which is strictly increasing and concave.

Substituting $\hat{\mu} = \sum_{i=1}^{n} \frac{\tau_i}{n}$ into the second equation, we obtain

$$\log(\hat{k}) - \Psi(\hat{k}) = \log(\hat{\mu}) - \frac{\sum_{i=1}^{n} \log \tau_i}{n} \tag{1.106}$$

which can be solved numerically, using as a starting point

$$\hat{k}_0 = \frac{1}{\mathbb{C}v^2} \tag{1.107}$$

Asymptotic confidence intervals can be obtained by

$$\begin{cases} \mathbb{V}ar[\hat{\mu}] = \frac{\mu^2}{nk} \\ \mathbb{V}ar[\hat{k}] = \frac{1}{n(\Psi'(k) - \frac{1}{k})} \\ \mathbb{C}ov(\hat{\mu}, \hat{k}) = 0 \end{cases} \tag{1.108}$$

1.1.5.3 Some Examples

In the following examples, we first report the classical, historical example of Von Bortkewitsch (1898) related to the horse kicks in the Prussian army, and then we add some case data sets, arising in different fields, to which the Poisson model could be applied. The first seven of them are cited and referenced in Lemaire (1995).

As a test of fit we use the classical Pearson χ^2 statistic with $\max(k) - p - 1$ degrees of freedom, where k denotes the number of groups and p denotes the number of estimated parameters. The best fit has minimum χ^2 and maximum p-value, i.e. the area at the right of the observed χ^2 statistics. Three (***) asterisks denote rejection of the Null Hypothesis at a confidence level (often well) below 1%.

In all examples the sample average $\bar{\mu}$ also provides the maximum likelihood estimate (MLE) of the Poisson rate parameter $\hat{\lambda}$. In some cases the parameters are estimated by the Method-of-Moments estimator (MME).

Example I.1 (Kicks). The following is the classic example introduced by Von Bortkewitsch (1898, p. 25), who applied the Poisson distribution to the number of deaths due to horse kicks in the ten corps of the Prussian army between 1875 and 1894 (20 years).

Von Bortkewitsch (1898), $n = 200$					
observed k's (with grouping)	observed freq. n_k	Poisson freq np_k (MLE)	NB freq np_k (MLE)	ZIP freq np_k (MME)	PIG freq np_k (MLE)
0	109	108.67	111.99	108.74	110.18
1	65	66.29	61.8	66.18	64.24
2	22	20.22	20	20.24	20.12
3+	4	4.82	6.22	4.84	5.47
$\bar{\mu} = 0.61$		$\chi_2^2 = 0.33$	$\chi_1^2 = 1.24$	$\chi_1^2 = 0.32$	$\chi_1^2 = 0.59$
$s^2 = 0.611$	p-value:	0.8506	0.2657	0.5706	0.4417

The Poisson model appears to be the best one. Data are compatible with pure homogeneous randomness.

Example I.2 (Wars). The example applies the Poisson distribution to a data set representing "the number of major international wars during each of the 432 years between 1500 and 1931".

Richardson (1945), $n = 432$					
observed k's (with grouping)	observed freq n_k	Poisson freq np_k (MLE)	NB freq np_k (MLE)	ZIP freq np_k (MME)	PIG freq np_k (MLE)
0	223	216.25	222.99	224.96	221.73
1	142	149.02	149.64	136.23	144.5
2	48	51.78	50.65	53.53	50.32
3+	19	14.33	17.31	17.28	15.46
$\bar{\mu} = 0.692$		$\chi_2^2 = 2.4$	$\chi_1^2 = 0.31$	$\chi_1^2 = 1.01$	$\chi_1^2 = 0.97$
$s^2 = 0.757$	p-value:	0.3016	0.5773	0.316	0.3245

All the models are accepted, but the Negative Binomial distribution performs better than all the other ones.

Example I.3 (Soccer). The example reports another application to sport data (soccer): here, the data show "the number of goals scored per (soccer) game, per team in the English First Division (now Premier League) in the 1967-1968 season".

Pollard et Al. (1977), $n = 924$					
observed k's (with grouping)	observed freq. n_k	Poisson freq np_k (MLE)	NB freq np_k (MLE)	ZIP freq np_k (MME)	PIG freq np_k (MLE)
0	225	203.71	225.25	242.94	247.93
1	293	308.01	297.25	265.08	288.67
2	224	232.86	215.05	220.29	197.42
3	114	117.36	112.84	122.05	105.4
4	41	44.36	47.99	50.72	49.12
5+	27	17.69	25.61	22.92	35.47
$\bar{\mu} = 1.512$		$\chi_4^2 = 8.54$	$\chi_3^2 = 1.54$	$\chi_3^2 = 7.45$	$\chi_3^2 = 9.83$
$s^2 = 1.739$	p-value:	0.0736	0.6728	0.0589	0.02

The Negative Binomial distribution provides by far the best model, but also the Poisson and ZIP distributions are not rejected as well.

Example I.4 (Hockey). This example (Benjamin et al., 1977) reports ice hockey scores, i.e. "the number of goals scored per game per team in the 1966-67 season of the National Hockey League".

Pollard et Al. (1977), $n = 420$					
observed k's (with grouping)	observed freq. n_k	Poisson freq np_k (MLE)	NB freq np_k (MLE)	ZIP freq np_k (MME)	PIG freq np_k (MLE)
0	29	21.35	27.46	41.57	60.79
1	71	63.61	68.77	52.71	86.96
2	82	94.75	91.59	83.47	78.92
3	89	94.09	86.17	88.12	60.02
4	65	70.07	64.23	69.77	42.32
5	45	41.75	40.33	44.19	28.94
6	24	20.73	22.18	23.32	19.58
7+	15	13.63	19.27	16.85	42.5
$\bar{\mu} = 2.979$		$\chi_6^2 = 6.86$	$\chi_5^2 = 2.9$	$\chi_5^2 = 10.74$	$\chi_5^2 = 73.5$
$s^2 = 3.539$	p-value:	0.3339	0.7154	0.0567	0***

The Poisson and the Negative Binomial distributions fit the data well; also the ZIP model is not rejected (at 5% confidence level).

Example I.5 (MTPL Belgium). Lemaire (1995, p. 25) reports data from a portfolio of automobile third-party liability contracts of a Belgian company, observed in 1976, and representative of the average market (the country claim frequency was about 10%), showing $n = 106,974$ units (sample mean $\hat{\mu} = 0.1011$ and sample variance $s^2 = 0.1074$).

Lemaire (1995), $n = 106,974$					
observed k's	observed freq n_k	Poisson freq np_k (MLE)	NB freq np_k (MLE)	ZIP freq np_k (MME)	PIG freq np_k (MME)
0	96,978	96,689	96,983	97,001	96,693
1	9,240	9,773	9,229	9,177	9,768
2	0.704	494	709	753	497
3	0.043	16.6	50.1	41.2	16.9
4+	0.009	0.4	3.6	1.7	0.44
$\bar{\mu} = 0.101$		$\chi_2^2 = 332$	$\chi_1^2 = 9.01$	$\chi_1^2 = 33.8$	$\chi_1^2 = 321$
$s^2 = 0.108$	p-value:	0***	0.0027***	0***	0***

All the models are heavily rejected.

The MTPL case is clearly an example of "*positive* contagion" (compare the expected frequency w.r.t. the Poisson) and an example of "*apparent* contagion" (see § 1.1.4) because the "contagion" is the effect of policyholders heterogeneity. The rejection of all models suggest that *risk classes*, as in the typical Bonus-Malus system, could be useful to model the population heterogeneity.

1.2 The Weighted Poisson Process of Order i

The Poisson process of order i, $N^{(i)}$, was introduced in 1984 and 1986 papers Philippou (1984), Charalambides (1986). The Poisson process of order i is also defined as a *weighted* sum of independent Poisson processes, i.e.

$$N^{(i)}(t) = \sum_{j=1}^{i} jN_j(t) = \sum_{j=1}^{i} \sum_{h=j}^{i} N_h(t) \tag{1.109}$$

where the processes $N_j(t)$ are i.i.d. Poisson processes, each weighted by j, and the process represents a flow of grouped events, i.e. whose arrivals are "grouped in packages".

Note that the processes are often assumed identically distributed, i.e. the arrival rate λ is unique.

The process can represent situations of multiple claims arrival or collective risks, possibly with heterogeneous (and decreasing) arrival rates $\lambda_1 > \lambda_2 > ... > \lambda_i$.

Its probability generating function - as obtained directly from the process definition – reads as:

$$\begin{aligned}
G^{(i)}(u,t) &= \\
&= \mathbb{E}\left[u^{N^{(i)}(t)}\right] = \mathbb{E}\left[u^{\sum_{j=1}^{i} jN_j}\right] = \\
&= \mathbb{E}\left[\prod_{j=1}^{i} u^{jN_j}\right] = \prod_{j=1}^{i} \mathbb{E}(u^j)^{N_j} = \\
&= \prod_{j=1}^{i} \mathbb{E}\left[\mathbb{E}[(u^j)^{N_j}|N_j]\right] = \\
&= \prod_{j=1}^{i}\sum_{h=0}^{\infty} (u^j)^h \cdot e^{-\lambda t}\frac{(\lambda t)^h}{h!} = \\
&= \prod_{j=1}^{i} e^{-\lambda t(1-u^j)} = e^{-i\lambda t}e^{\lambda t \sum_{j=1}^{i} u^j} = \\
&= e^{-i\lambda t}e^{\lambda t\left(u\frac{1-u^i}{1-u}\right)} = \\
&= e^{-i\lambda t\left(1-\frac{u}{i}\frac{1-u^i}{1-u}\right)}.
\end{aligned} \tag{1.110}$$

Its probability generating function can be derived as well from a representation of the process as a compound Poisson process with discrete-Uniform compounding distribution, i.e. $N^{(i)}(t) = X_1 + X_2 + ... + X_{N_{i\lambda}(t)}$, where $N_{i\lambda}(t)$ is a Poisson process of rate $i\lambda$, and X_h are Uniform r.v.'s over the set of the first i integers.

Hence, it is obtained as

$$
\begin{aligned}
G^{(i)}(u,t) &= \mathbb{E}\left[u^{N^{(i)}(t)}\right] = \mathbb{E}\left[u^{\sum_{h=0}^{N_{i\lambda}(t)} X_h}\right] = \\
&= \mathbb{E}\left[\mathbb{E}\left[u^{\sum_{h=0}^{N_{i\lambda}(t)}} | N_{i\lambda}(t)\right]\right] = \mathbb{E}\left[\mathbb{E}\left(u^{X_h}\right)^{N_{i\lambda}(t)}\right] = \\
&= \sum_{h=0}^{\infty}\left(\mathbb{E}\left[u^{X_h}\right]\right)^h \cdot e^{-i\lambda t}\frac{(i\lambda t)^h}{h!} = e^{-i\lambda t}e^{i\lambda t \cdot \mathbb{E}u^X} = \\
&= e^{-i\lambda t(1-\psi_X(u))}
\end{aligned}
\tag{1.111}
$$

where the p.g.f. of the discrete-Uniform is $\psi_X(u) = \frac{u}{i}\frac{1-u^i}{1-u}$.

Remark 1.7.
The p.g.f. of the discrete-Uniform r.v.'s X_h, given that the probability mass function of the discrete-Uniform distribution over the set of the *first* i integers is $\Pr\{X = k\} = \frac{1}{i}$, is obtained as $\mathbb{E}[u^{X_h}] = \sum_{k=1}^{i}\frac{1}{i}u^k = \frac{u}{i}\frac{1-u^i}{1-u}$, as a result of the truncated sum of the geometric series of ratio $u < 1$, starting from 1.

An explicit expression of the state probabilities can be found by the coefficients of the p.g.f. (see Kostadinova and Minkova (2013) for its expression for $k > i$)

$$
\begin{cases}
p_0^{(i)}(t) = e^{-i\lambda t} \\
p_k^{(i)}(t) = e^{-i\lambda t}\sum\limits_{h=1}^{k}\binom{k-1}{h-1}\frac{(\lambda t)^h}{h!} \qquad k = 1, 2, ..., i
\end{cases}
\tag{1.112}
$$

One can note that the maximum number of events for a given i over an interval dt corresponds to the sum of the first i integers, i.e. the maximal jump size is $\frac{i(i+1)}{2}$, which can be substantially larger than 1.

Its mean and variance (see Fig. 1.4 and Fig. 1.5 respectively) follow the expression for the sum of the first i integers, namely

$$
\begin{cases}
\mathbb{E}[N^{(i)}(t)] = \lambda t\sum\limits_{j=1}^{i} j = \frac{i(i+1)}{2}\lambda t \\
\mathbb{V}ar[N^{(i)}(t)] = \lambda t\sum\limits_{j=1}^{i} j^2 = \frac{i(i+1)(2i+1)}{6}\lambda t
\end{cases}
\tag{1.113}
$$

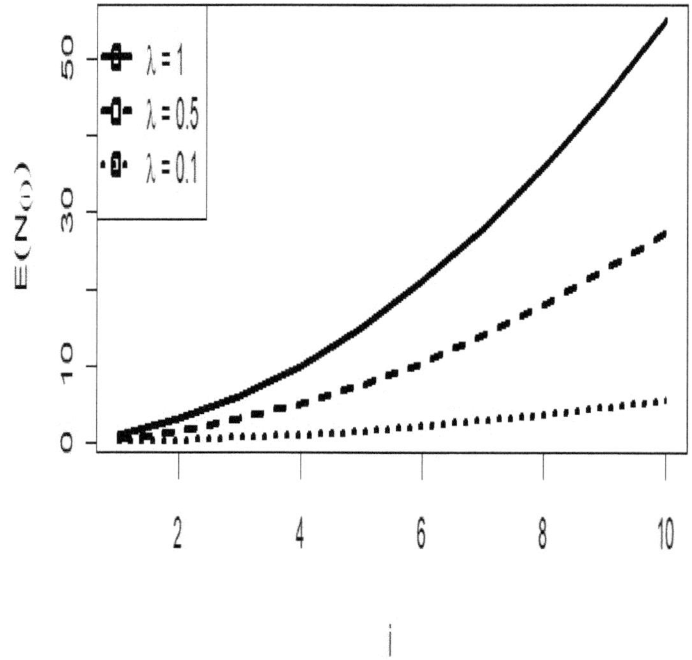

FIGURE 1.4
Mean of the weighted Poisson process w.r.t. the process index i ($t = 1$).

The process is overdispersed, that is, the ratio of variance over the mean, also known as Fisher index is greater than one, i.e. $FI = 1 + \frac{2}{3}(i-1) > 1$, which in turn increases with the order i of the process.

A second representation is described in Kostadinova and Minkova (2013), who illustrate the (linear) pure-birth (see § 2.2) representation of the weighted Poisson process, where

$$\begin{cases} \Pr\{N^{(i)}(t+dt) = h+k | N^{(i)}(t) = h\} = \lambda dt + o(dt), k = 1, 2, ..., i \\ \qquad\qquad\qquad\qquad\qquad\qquad\qquad\qquad\qquad \text{(multiple jumps)} \\ \Pr\{N^{(i)}(t+dt) = h+k | N^{(i)}(t) = h\} = 1 - i\lambda dt + o(dt), k = 0; \end{cases} \tag{1.114}$$

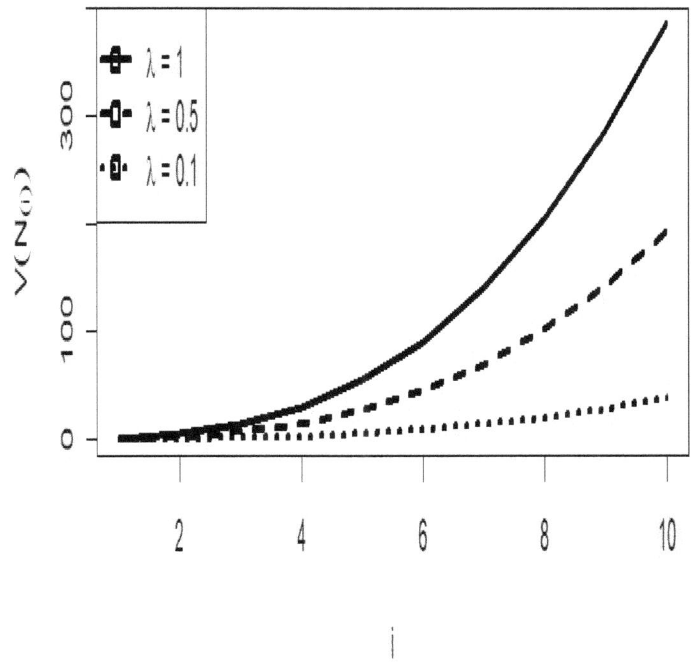

FIGURE 1.5
Variance of the weighted Poisson process w.r.t. the process index i ($t = 1$).

implicitly, for jumps of amplitude greater than i, it holds that $\Pr\{N^{(i)}(dt) \geq i + 1\} \sim o(dt)$.

Alternatively, a more compact notation is (see Maheshwari and Orsingher, 2019)

$$\begin{cases} \Pr\{dN^{(i)}(t) = k\} = \lambda dt & 1 \leq k \leq i \\ \Pr\{dN^{(i)}(t) = 0\} = 1 - i\lambda dt \end{cases} \tag{1.115}$$

1.2.1 Estimation of Poisson Process of Order i

In the following section Method-of-Moments and Maximum Likelihood estimators for the Poisson process of order i are illustrated.

1.2.1.1 Estimators

Method-of-Moments Estimator. Aki and Hirano (1989) provide straightforward expressions for the Method-of-Moments Estimator (MME) of the process. Let us set $t = 1$ to deal with the r.v. (and not the process), and let us suppose to observe a set of n independent observations $\underline{\mathbf{x}} \equiv \{x_1, x_2, ..., x_h, ..., x_n\}$ compatible with the multiple arrival structure of the process, then, since the mean and the variance of the Poisson of order i r.v. are $\lambda \cdot \frac{i(i+1)}{2}$ and $\lambda \cdot \frac{i(i+1)(2i+1)}{6}$, respectively, the MME $\hat{\lambda}_{MME}$ is

$$\hat{\lambda} = \frac{2}{i(i+1)}\bar{x}_n \tag{1.116}$$

with (*exact*) variance given by

$$\sigma_n^2 = \frac{2(2i+1)}{3i(i+1)}\frac{\lambda}{n} \tag{1.117}$$

which tends to zero as the sample size n increases (the MME estimator is unbiased).

Aki and Hirano (1989) give the *asymptotic* (i.e. independent of the sample size n) variance:

$$\sigma_\infty^2 = \frac{2(i+1)}{3i(i+1)}\frac{1}{\lambda} \tag{1.118}$$

Maximum Likelihood Estimator. Aki and Hirano (1989) first studied the ML parameter estimation for the Poisson process of order i, in the case of a homogeneous parameter λ (as opposed to a set of parameters λ_h, $h = 1, 2, ..., i$).

They exploit the linear structure of the probability mass function $p_k^{(i)}(\lambda)$ (PMF) to apply the first and second derivatives necessary to pursue a likelihood analysis.

In fact, provided the following expression of the PMF ($t = 1$)

$$\begin{cases} p_0^{(i)}(\lambda) = e^{-i\lambda} \\ p_k^{(i)}(\lambda) = \frac{\lambda}{k} \sum_h^{k \wedge i} h \cdot p_{k-h}^{(i)}(\lambda) \end{cases} \tag{1.119}$$

Remark 1.8a. Generalized Panjer recursion

Note that, due to the Compound Poisson structure of the process of order i, i.e. $\sum_{h=1}^{N_{i\lambda}(t)} X_h$, with summands X_h distributed as a discrete-Uniform distribution over the first i (positive) integers $k \leq i$ (see note 2 above), formula (1.119) arises from the Panjer recursion ($p_k = (a + \frac{b}{k}h) \cdot f_h \cdot p_{k-h}$, $k = 1, 2, ...$; $a, b \geq 0$) for Compound Poisson distributions defined on positive integers (Panjer, 1981, eq. 5), i.e. when the compounding distribution $F(x)$ of X is discrete, i.e.:

$$p_k = \sum_h^k (a + \frac{b}{k}h) f_h p_{k-h} \tag{1.120}$$

Then, for the Weighted Poisson process of order i, $a = 0$, $b = i \cdot \lambda$ and the above recursion holds, i.e. ($f_h \equiv \frac{1}{i}$).

$$p_k^{(i)} = \frac{\lambda}{k} \sum_h^{k \wedge i} h p_{k-h}^{(i)}. \tag{1.121}$$

Note that here the lagged probabilities p_{k-h} obtain as a result of the convolution behind Panjer recursion formulae; their probabilistic intuition is that the process allows for multiple jumps (over any infinitesimal time interval dt). Note finally that Andelson (1966) obtained Panjer's recursion for Compound Poisson distributions limited to discrete claim amounts and using p.g.f.'s.

Remark 1.8b. Simple Panjer recursion
The simple recursion, for suitable parameters a, b

$$p_k = \left(a + \frac{b}{k}\right) \cdot p_{k-1} \qquad k \geq 1 \tag{1.122}$$

defines a set of discrete distributions.

- For $a = -\frac{p}{1-p}$ and $b = (n+1)\frac{p}{1-p}$, $p_0 = (1-p)^n$, we have the Binomial(p, n) distribution;
- for $a = 0$ and $b = \lambda$, $p_0 = e^{-\lambda}$, we have the Poisson(λ) distribution;
- for $a = 1-p$ and $b = (1-p)(r-1)$, $p_r = p^r \quad (k > r)$, we have the Negative Binomial(p, r) distribution, whose special case $r = 1$ is the Geometric(p) distribution;
- for $a = p$ and $b = -p$, $p_r = -\frac{1}{\ln(1-p)}p$, we have the logarithmic distribution.

For Panjer *distributions*

$$\begin{cases} \mathbb{E}[X] = \frac{a+b}{1-a} \\ \mathbb{V}ar[X] = \frac{a+b}{(1-a)^2} \end{cases} \tag{1.123}$$

Aki and Hirano (1989) show that the MLE problem is given by the equation

$$L'(\lambda) = \sum_{h=1}^{n} \frac{d}{d\lambda} \ln\{p^{(i)}(\lambda; x_h)\} = \sum_{i=1}^{n} \frac{\frac{d}{d\lambda} p^{(i)}(\lambda; x_h)}{p^{(i)}(\lambda; x_h)} \equiv G(\lambda) = 0 \tag{1.124}$$

to be solved numerically, by the Newton-Raphson recursion, whose generic j-th step is

$$\lambda_{j+1} = \lambda_j - \frac{G(\lambda_j)}{G'(\lambda_j)} \tag{1.125}$$

where $G'(\lambda) = \sum_{h=1}^{n} \frac{p'' \cdot p - p'^2}{(p)^2}$.

The expressions of the first and second derivatives of the probabilities $p_k^{(i)}(\lambda)$ are, respectively:

$$\begin{cases} p_0'^{(i)}(\lambda) = -ie^{-i\lambda} \\ p_k'^{(i)}(\lambda) = \frac{1}{k}\sum_{h=1}^{k\wedge i} h \cdot p_{k-h}(\lambda) + \frac{\lambda}{k}\sum_{h=1}^{k\wedge i} h \cdot p'_{k-h}(\lambda) \end{cases} \tag{1.126}$$

and

$$\begin{cases} p''_0{}^{(i)}(\lambda) = i^2e^{-i\lambda} \\ p_k'^{(i)}(\lambda) = \frac{2}{k}\sum_{h=1}^{k\wedge i} h \cdot p'_{k-h}(\lambda) + \frac{\lambda}{k}\sum_{h=1}^{k\wedge i} h \cdot p''_{k-h}(\lambda) \end{cases} \tag{1.127}$$

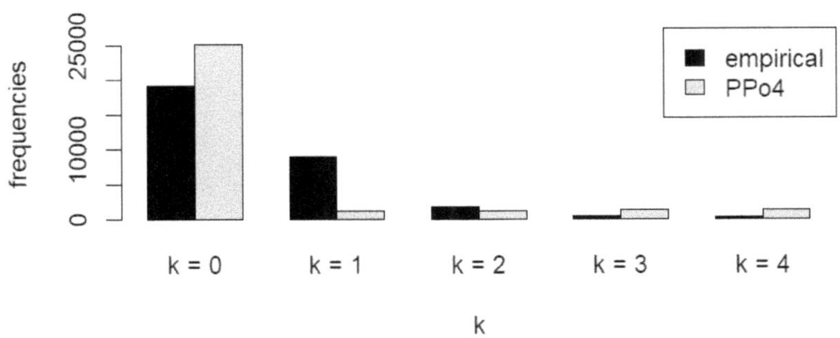

FIGURE 1.6
Bar plot showing empirical data and fitted values using a Weighted Poisson of order 4.

1.3 Skellam Process: The Difference Between Two Poisson Processes

The process obtained as the difference of two Poisson processes was studied first in 1946 by Skellam. It is defined as $S(t) = N_1(t) - N_2(t)$, where $N_1(t)$, $N_2(t)$, $t > 0$, are independent Poisson processes, with rates $\lambda_1 > 0$, $\lambda_2 > 0$, and take integer values (i.e., both positive or negative). Jumps have unitary size.

The probability mass function of $S(t)$ turns out to be of the form (see Fig. 1.4)

$$p_k(t) \equiv \Pr\{S(t) = k\} = e^{-(\lambda_1+\lambda_2)t} \left(\frac{\lambda_1}{\lambda_2}\right)^{\frac{k}{2}} I_{|k|}(2\sqrt{\lambda_1\lambda_2 t}), \quad k \in \mathbb{Z} \tag{1.128}$$

where $I_k(x)$ is the modified Bessel function of the first kind, defined as $I_k(x) = \sum_{n=0}^{\infty} \frac{(\frac{x}{2})^{2n+k}}{n!\Gamma(n+k+1)}$. Of course, if $\lambda_1 = \lambda_2 = \lambda$ the distribution $\{p_k, k \geq 0\}$ is symmetric and reads $p_k(t) = e^{-2\lambda t} I_{|k|}(2\lambda t)$; positive (negative) skewness corresponds to respectively $\lambda_1 > \lambda_2$ $(\lambda_1 < \lambda_2)$. The mean, variance, covariance and p.g.f. of the distribution are respectively

$$\begin{cases} \mathbb{E}[S(t)] = (\lambda_1 - \lambda_2)t, \\ \mathbb{V}ar[S(t)] = (\lambda_1 + \lambda_2)t, \\ \mathbb{C}ov(S(t), S(s)) = (\lambda_1 + \lambda_2)\min(s,t), \\ G_S(u) = G_{N_1}(u)G_{N_2}(\frac{1}{u}) \end{cases} \tag{1.129}$$

Note that the process is overdispersed.

Theorem 1.11. *The state probabilities of the Skellam process $p_k(t)$ are governed by the difference-differential equation*

$$\frac{dp_k(t)}{dt} \equiv -(\lambda_1 p_k(t) - \lambda_1 p_{k-1}(t)) + (\lambda_2 p_{k+1}(t) - \lambda_2 p_k(t)) \tag{1.130}$$

with initial conditions $p_0(0) = 1$, $p_k(0) = 0$, $k \in \mathbb{Z} \setminus 0$.

Proof. Clearly

$$\begin{aligned} &p_k(t + dt) = \\ &= p_k(t)(1 - \lambda_1 dt)(1 - \lambda_2 dt) + \\ &+ p_{k-1}(t)\lambda_1 dt(1 - \lambda_2 dt) + p_{k+1}(t)(1 - \lambda_1 dt)\lambda_2 dt = \\ &= p_k(t) - p_k(t)\lambda_1 dt - p_k(t)\lambda_2 dt + p_{k-1}(t)\lambda_1 dt + p_{k+1}(t)\lambda_2 dt \end{aligned} \tag{1.131}$$

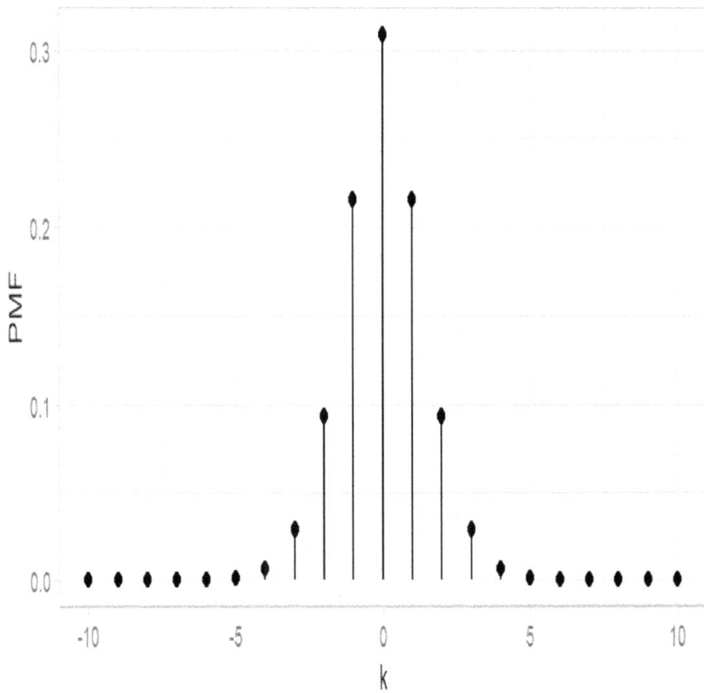

FIGURE 1.7
Probability mass function of the Skellam process with symmetry ($\lambda_1 = \lambda_2 = 1$) w.r.t. ($t = 1$).

thus

$$\frac{dp_k(t)}{dt} = -p_k(t)(\lambda_1 + \lambda_2) + \lambda_1 p_{k-1}(t) + \lambda_2 p_{k+1}(t) \tag{1.132}$$

is obtained.

□

The sample paths of the Skellam process have unitary upwards and downwards jumps: as such, the Skellam distribution is suited for modelling data representing differences (e.g. soccer scorings) or increments (e.g. up and down price movements). An actuarial example is the application to experience (bonus-malus) rating in motor-vehicle insurance.

The moment generating function of the Skellam process reads as (see Buchak and Sakhno, 2019)

$$M(u;t) = \mathbb{E}[e^{uS(t)}] = \exp\left\{-\left[\lambda_1(1-e^{u}) + \lambda_2(1-e^{-u})\right]t\right\}, \quad |u| \leq 1 \tag{1.133}$$

1.3.1 Estimation of the Skellam Process

This section reports the Method-of-Moments estimators for the two intensity parameters of the Skellam process.

1.3.1.1 Estimators

The expected value of the Skellam process is given by

$$\mathbb{E}[S(t)] = (\lambda_1 - \lambda_2)t \tag{1.134}$$

while its variance is given by

$$\mathbb{V}ar[S(t)] = (\lambda_1 + \lambda_2)t \tag{1.135}$$

Given observations and their sample mean m and sample variance s^2, and assuming $t = 1$, the method-of-moments estimators solve the equations

$$\begin{cases} \lambda_1 - \lambda_2 = m \\ \lambda_1 + \lambda_2 = s^2 \end{cases} \tag{1.136}$$

whose solutions are

$$\begin{cases} \hat{\lambda}_1 = \frac{s^2+m}{2} \\ \hat{\lambda}_2 = \frac{s^2-m}{2} \end{cases} \tag{1.137}$$

1.3.1.2 Some Examples

The Skellam distribution is often applied to soccer games, because the record of the difference in scores of a match reproduces a sample from the Skellam distribution.

Example (Football scores). As an example, we consider the Italian national league (Serie A), and to fix ideas, we chose the first and the last classified teams in the season 2021-2022, i.e. the teams of "Milan" and "Venezia". For example, the match "Juventus-Milan" 1-1 produces a goal difference $x = 0$; the match "Bologna-Venezia 0-1" produces a goal difference $x = -1$, and so on and so forth.

Over 38 matches, data[3] report the score differences x_i, $i = 1, 2, ..., 38$, with sample mean m and sample variance s^2:

$$\begin{cases} m_{Milan} = 0.97; s^2_{Milan} = 1.49 \\ m_{Venezia} = -1.08; s^2_{Venezia} = 1.37 \end{cases}$$

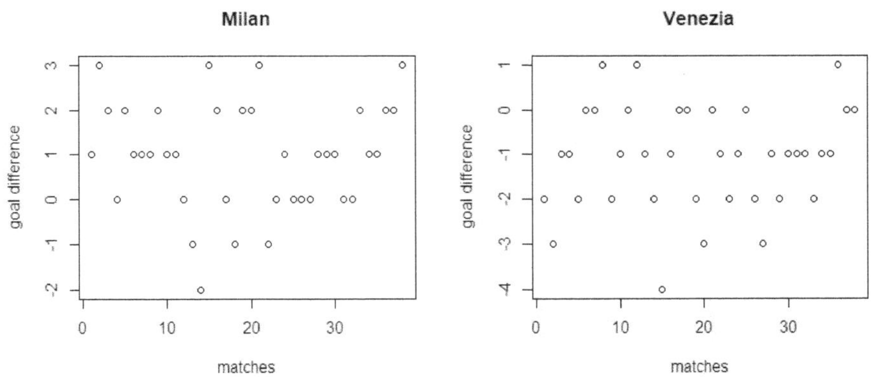

FIGURE 1.8
Goal difference over the 38 matches for "Milan" and "Venezia" football clubs.

Then the parameters estimates are

$$\begin{cases} \lambda_{1,+} = 1.23 \\ \lambda_{1,-} = 0.26 \\ \lambda_{2,+} = 0.15 \\ \lambda_{2,-} = 1.23 \end{cases} \tag{1.138}$$

[3]Source: https://www.legaseriea.it/it/serie-a/statistiche

One can notice how the λ_+ parameter (related to the positive jumps) is higher for the "Milan" club (1st classified), and vice versa for the "Venezia" club (last classified).

Milan (2021-2022), $n = 38$		
observed k's	observed freq n_k	Skellam freq np_k (MME)
-2	1	0.32
-1	3	2.6
0	9	11.5
+1	12	12.3
+2	9	7.2
+3	4	2.9
$\bar{\mu} = 0.97$	$\lambda_+ = 1.23$	χ^2_3 1.0445
$s^2 = 1.49$	$\lambda_- = 0.26$	p-value $= 0.9589$

Venezia (2021-2022), $n = 38$		
observed k's	observed freq n_k	Skellam freq np_k (MME)
-4	1	0.9
-3	3	3.1
-2	9	7.7
-1	13	12.9
0	9	11.4
+1	3	1.6
$\bar{\mu} = -1.08$	$\lambda_+ = 0.15$	$\chi^2_3 = 0.8148$
$s^2 = 1.37$	$\lambda_- = 1.23$	p-value $= 0.9761$

The Skellam distribution fits the goal difference data properly, both for the first (Milan) and last (Venezia) team of the championship.

1.4 The Time Non-homogeneous Poisson Process

A generalisation of the Poisson process is obtained by assuming that the rate function $\lambda = \lambda(t)$ depends on time t.

The definition of a homogeneous Poisson process can be rewritten as:

1. $\Pr\{N(0)=0\}=1$;
2. $\Pr\{N(t,t+dt]=1\}=\lambda(t)dt+o(dt)$, $\lambda(t)>0$, $t>0$;
3. $\Pr\{N(t,t+dt]=0\}=1-\lambda(t)dt+o(dt)$;
4. $\Pr\{N(t,t+dt]>1\}=o(dt)$;
5. The process has independent increments in the sense that if $0=t_0<t_1<\ldots<t_n<t$, the random variables $N(t_j)-N(t_{j-1})$, $1\le j\le n$, are independent.

Equation (1.1) must be replaced by:

$$\frac{dp_k(t)}{dt}=-\lambda(t)p_k(t)+\lambda(t)p_{k-1}(t), \qquad k\ge 0, \tag{1.139}$$

with the initial conditions

$$p_k(0)=\begin{cases}1 & k=0\\ 0 & k>0\end{cases} \tag{1.140}$$

and $p_{-1}(t)=0$.

The probability generating function $G(u,t)$, $|u|\le 1$, $t>0$, emerges by solving the following partial differential equation:

$$\begin{cases}\frac{\partial}{\partial t}G(u,t)=\lambda(t)(u-1)G(u,t),\\ G(u,0)=1.\end{cases} \tag{1.141}$$

From (1.141) it is clear that

$$G(u,t)=e^{-(1-u)\int_0^t\lambda(s)ds}= \\ =e^{-\int_0^t\lambda(s)ds}\left[\sum_{k=0}^{\infty}\frac{\left[\int_0^t\lambda(s)ds\right]^k}{k!}u^k\right], \tag{1.142}$$

so that the probability distribution of the non-homogeneous Poisson process becomes

$$p_k(t)=e^{-\int_0^t\lambda(s)ds}\frac{\left[\int_0^t\lambda(s)ds\right]^k}{k!}, \qquad k\ge 0. \tag{1.143}$$

The time integral of the intensity function over the interval $(0, T]$ be denoted by

$$\Lambda(T) = \int_0^T \lambda(t)dt \tag{1.144}$$

and it represents the (time-varying) expected value, given that

$$\Lambda(T) = \mathbb{E}[N(T)] = \int_0^T \lambda(t)dt \tag{1.145}$$

In fact,

$$\begin{aligned} \mathbb{E}\left[N(t)\right] &= \sum_{k=0}^{\infty} k e^{-\int_0^t \lambda(s)ds} \frac{\left(\int_0^t \lambda(s)ds\right)^k}{k!} = \\ &= e^{-\int_0^t \lambda(s)ds} \sum_{k=1}^{\infty} \frac{\left(\int_0^t \lambda(s)ds\right)^{k-1}}{(k-1)!} \int_0^t \lambda(s)ds = \int_0^t \lambda(s)ds \end{aligned} \tag{1.146}$$

The second moment reads

$$\begin{aligned} \mathbb{E}\left[N^2(t)\right] &= \sum_{k=0}^{\infty} k^2 e^{-\int_0^t \lambda(s)ds} \left(\int_0^t \lambda(s)ds\right) = \\ &= e^{-\int_0^t \lambda(s)ds} \sum_{k=1}^{\infty} k \frac{\left(\int_0^t \lambda(s)ds\right)^{k-1}}{(k-1)!} \int_0^t \lambda(s)ds = \\ &= e^{-\int_0^t \lambda(s)ds} \int_0^t \lambda(s)ds \sum_{k'=0}^{\infty} (1+k') \frac{\left(\int_0^t \lambda(s)ds\right)^{k'}}{k'!} = \\ &= e^{-\int_0^t \lambda(s)ds} \int_0^t \lambda(s)ds \left(e^{\int_0^t \lambda(s)ds} + \int_0^t \lambda(s)ds\right) = \\ &= \int_0^t \lambda(s)ds \left(\int_0^t \lambda(s)ds + 1\right) \end{aligned} \tag{1.147}$$

and, hence, the variance

$$\mathbb{V}ar\left[N(t)\right] = \int_0^t \lambda(s)ds \left(\int_0^t \lambda(s)ds + 1\right) - \left(\int_0^t \lambda(s)ds\right)^2 = \\ = \int_0^t \lambda(s)ds \tag{1.148}$$

Clearly,

$$\Pr\{N(T) = n\} = \Lambda(T)^n \frac{e^{-\Lambda(T)}}{n!} \qquad \forall n \geq 0 \tag{1.149}$$

Note that the non-homogeneous Poisson process can be seen as a homogeneous Poisson process with a time-change and unit intensity:

$$N(t) = N^{\star}(\Lambda(t)) \tag{1.150}$$

In general, we have, for $s < t$, that if $N(t)$ is non-homogeneous with intensity $\Lambda(t)$, then $N(t) - N(s)$ is non-homogeneous with intensity $\Lambda(s,t)$, that is

$$\Pr\{N(t) - N(s) = k\} = \frac{\Lambda(s,t)^k e^{-\Lambda(s,t)}}{k!} \tag{1.151}$$

with $\Lambda(s,t) = \int_s^t \lambda(u)du = \Lambda(t) - \Lambda(s)$ and we have that the increments $N(t) - N(s)$ are no longer stationary.

Theorem 1.12. *Given a Poisson process with parameter λ, $N(t)$, its events are classified of type i using a distribution $p_i(s)$, $i = 1, ..., K$, $\sum_{i=1}^K p_i(s) = 1$, $s < t$. Then $N_i(t)$, $i = 1, ..., K$ are independent non-homogeneous Poisson processes with mean*

$$\mathbb{E}[N_i(t)] = \lambda \int_0^t p_i(s)ds \equiv \Lambda_i(t) \tag{1.152}$$

and

$$Pr\{N_i(t) = n_i\} = e^{-\lambda\Lambda_i(t)} \frac{(\lambda\Lambda_i(t))^{n_i}}{n_i!} \tag{1.153}$$

Proof. See Ross (2024), Th. 5.6, p. 336. □

Example (Logistic intensity function). Specific growth models can be constructed by modelling the density function so as to obtain a non-homogeneous Poisson process $\hat{N}(t)$ with a specific pattern: e.g. one such example (see Di Crescenzo, 2019) is obtained by modeling the intensity rate $\lambda(t)$ with a (decreasing) *logistic* function, i.e.

$$\lambda(t) = \frac{r(C - n_0)}{C - n_0 + n_0 e^{rt}} \tag{1.154}$$

where C is the so-called *capacity* or maximum point, n_0 is the starting point or minimum point and r is the growth rate.

One can note that as time grows and the rate function decreases, then, correspondingly, the logistic curve reaches a "plateau" (its slope also decreases). Hence, correspondingly, the (non-homogeneous) Poisson process jumps at more and more distant time intervals (see Fig. 1.9).

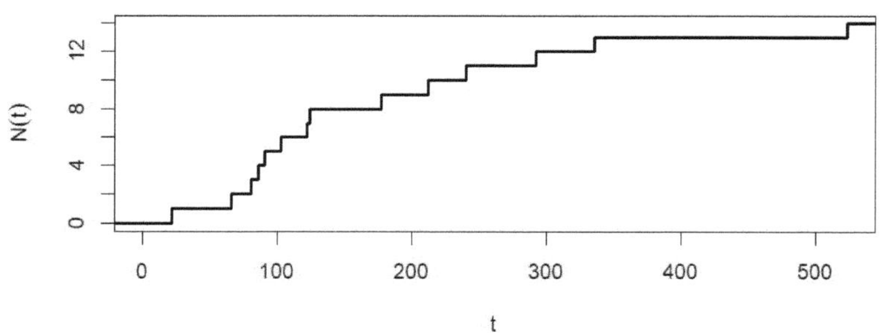

FIGURE 1.9
Sample path of a non-homogeneous Poisson process with logistic intensity function.

1.4.1 Estimation of the Non-homogeneous Poisson Process

Likelihood-based inference naturally generalizes to the inhomogeneous case. The likelihood takes a tractable form depending on the intensity function.

The probability of one event between two subsequent time points t_i and t_{i+1} (i.e. the probability of the next event) is $\lambda(t_{i+1})e^{-(\Lambda(t_{i+1})-\Lambda(t_i))}$.

Given the time-sample $(0, T]$, we assume that n events are observed, $N(T) = n$, at the time points $\underline{t} \equiv (t_1, t_2, ..., t_i, ..., t_n)$. Then the joint probability density of the n arrival times $0 < t_1 < ... < t_i < ... < t_n < T$ is given by (see Cox and Lewis, 1966, p. 45)

$$\begin{aligned} &\lambda(t_1)e^{-(\Lambda(t_1)-\Lambda(0))} \cdot \lambda(t_2)e^{-(\Lambda(t_2)-\Lambda(t_1))} \cdots \\ &\quad \cdots \lambda(t_n)e^{-(\Lambda(t_n)-\Lambda(t_{n-1})} \cdot e^{-(\Lambda(T)-\Lambda(t_n))} = \\ &\qquad = e^{-\Lambda(T)} \prod_{i=1}^{n} \lambda(t_i), \qquad \Lambda(0) = 0. \end{aligned} \tag{1.155}$$

Note that the last factor $e^{-(\Lambda(T)-\Lambda(t_n))}$ stands for the probability of no events in $(t_n, T]$.

Example (Exponential intensity function, Cox and Lewis, 1966, p. 45). Let $\lambda(t) = e^{\alpha+\beta t}$, $t > 0$. Then

$$\Lambda(T) = \int_0^T e^{\alpha+\beta t} dt = \frac{e^{\alpha}}{\beta}(e^{\beta T} - 1) \tag{1.156}$$

and the likelihood function is

$$L(\alpha, \beta, \underline{t}) = \exp\left(n\alpha + \beta \sum_{i=1}^{n} t_i - \frac{e^{\alpha}}{\beta}(e^{\beta T} - 1)\right) \tag{1.157}$$

Taking logarithms and deriving w.r.t. the parameters, one obtains

$$\begin{cases} \frac{\partial \log L}{\partial \alpha} = n - \frac{e^{\alpha}}{\beta}\left(e^{\beta T} - 1\right) = 0 \\ \frac{\partial \log L}{\partial \beta} = \sum_{i=1}^{n} t_i - e^{\alpha} \frac{Te^{\beta T}\beta - (e^{\beta T} - 1)}{\beta^2} = 0 \end{cases} \tag{1.158}$$

so that

$$\alpha_{MLE} = \log\left(\frac{n\beta}{e^{\beta T} - 1}\right) \tag{1.159}$$

and β_{MLE} is a solution of

$$\sum_{i=1}^{n} t_i - n\left(\frac{Te^{\beta T}}{e^{\beta T}-1} - \frac{1}{\beta}\right) = 0 \tag{1.160}$$

Note that for $\beta \approx 0$ the intensity is similar to a linear trend, i.e.

$$\lambda(t) \approx e^{\alpha}(1+\beta t) \tag{1.161}$$

From (1.157) we see that n and $\sum_{i=1}^{n} t_i$ are sufficient statistics for the parameters.

Moreover, given β, n is sufficient for α.

Therefore, the likelihood for β is the conditional density of $\sum_{i=1}^{n} t_i$ given n, i.e.

$$\begin{aligned} L\left(\beta, \sum_{i=1}^{n} t_i \,\middle|\, n\right) &= \frac{L(\alpha,\beta;\underline{t})}{\Pr\{N(T)=n\}} = \\ = \frac{\exp\{n\alpha + \beta\sum_{i=1}^{n} t_i - \frac{e^{\alpha}}{\beta}(e^{\beta T}-1)\}}{\left(\frac{e^{\alpha}(e^{\beta T}-1)}{\beta}\right)^{n} \frac{1}{n!}\exp\{-\frac{e^{\alpha}}{\beta}(e^{\beta T}-1)\}} &= \\ = \frac{n!\beta^{n}\exp\{n\alpha + \beta\sum_{i=1}^{n} t_i\}}{(e^{\alpha}(e^{\beta T}-1))^{n}} = \frac{n!\beta^{n}\exp(\beta\sum_{i=1}^{n} t_i)}{(e^{\beta T}-1)^{n}} \end{aligned} \tag{1.162}$$

Note also that this is the density of an ordered random sample of n values from the truncated Exponential distribution with density

$$\frac{\beta e^{\beta t}}{e^{\beta T}-1}, \qquad 0 \le t \le T \tag{1.163}$$

For $\beta = 0$ the Poisson process is homogeneous, the density is Uniform and (1.19) is recovered.

Taking logarithms one obtains

$$\log\left(L\left(\sum_{i=1}^{n} t_i, \beta \,\middle|\, n\right)\right) = n\log(\beta) - n\log(e^{\beta T}-1) + \beta\sum_{i=1}^{n} t_i + \log n! \tag{1.164}$$

and deriving w.r.t. $\beta \neq 0$

$$\frac{\partial \log(L)}{\partial \beta} = \frac{n}{\beta} - \frac{nT}{1 - e^{-\beta T}} + \sum_{i=1}^{n} t_i = 0 \tag{1.165}$$

as seen above in (1.160).

The information function is

$$I(\beta) = -\mathbb{E}\left[\frac{\partial^2 \log(L)}{\partial \beta^2}\right] = n\left(\frac{1}{\beta^2} - \frac{T^2 e^{-\beta T}}{(1 - e^{-\beta T})^2}\right) \tag{1.166}$$

and asymptotically, as $n \to \infty$, for testing $\beta = \beta_0$

$$\frac{\partial \log(L(\beta_0))}{\sqrt{I(\beta_0)}} \sim N(0,1) \tag{1.167}$$

To test the null hypothesis $\beta = 0$, we use the fact that $\sum_{i=1}^{n} t_i$ is the sum of Uniform distributions over $[0, T]$. It has therefore the Irwin-Hall distribution which is asymptotically Normal

$$\sum_{i=1}^{n} t_i \sim N\left(\frac{nT}{2}, \frac{nT^2}{12}\right) \tag{1.168}$$

The test for $\beta = 0$ is therefore, for $n \to \infty$

$$\frac{\frac{\sum_{i=1}^{n} t_i}{n} - \frac{T}{2}}{T\sqrt{\frac{1}{12n}}} \sim N(0,1) \tag{1.169}$$

Note that if T is an arrival time $T = t_n$, in the above formula n should be replaced by $n-1$ and the test is

$$\frac{\frac{\sum_{i=1}^{n} t_i}{n-1} - \frac{T}{2}}{T\sqrt{\frac{1}{12(n-1)}}} \sim N(0,1) \tag{1.170}$$

1.4.2 Doubly Stochastic (or Conditional) Poisson Process

Let $N(t)$ be a point process with random intensity $\tilde{\lambda}$ with distribution $G(\lambda)$ in $(0, \infty)$. Conditioning on $\tilde{\lambda} = \lambda$, $N(t)$ is a Poisson process.

Therefore

$$\begin{aligned}\Pr\{N(t)-N(s)=n\} &= \\ =\int_0^\infty \Pr\{N(t)-N(s)=n \mid \tilde{\lambda}=\lambda\}dG(\lambda) &= \\ =\int_0^\infty e^{-\lambda(t-s)}\frac{(\lambda(t-s))^n}{n!}g(\lambda)d\lambda\end{aligned} \tag{1.171}$$

so that the process has stationary but non-independent increments.

We have, using the conditional variance formula, that

$$\begin{cases}\mathbb{E}[N(t)]=\mathbb{E}[\tilde{\lambda}]t \\ \mathbb{V}ar[N(t)]=t\mathbb{E}[\tilde{\lambda}]+t^2\mathbb{V}ar[\tilde{\lambda}]\end{cases} \tag{1.172}$$

The conditional distribution of $\left(\tilde{\lambda} \mid N(t)=n\right)$ is obtained as

$$\begin{aligned}\Pr\{\tilde{\lambda}\le x \mid N(t)=n\} = \frac{\Pr\{\tilde{\lambda}\le x, N(t)=n\}}{\Pr\{N(t)=n\}} &= \\ =\int_0^\infty \Pr\{\tilde{\lambda}\le x, N(t)=n \mid \tilde{\lambda}=\lambda\}g(\lambda)d\lambda &= \\ =\int_0^x \frac{\Pr\{N(t)=n \mid \tilde{\lambda}=\lambda\}g(\lambda)d\lambda}{\Pr\{N(t)=n\}} &= \\ =\frac{\int_0^x e^{-\lambda t}(\lambda t)^n g(\lambda)d\lambda}{\int_0^\infty e^{-\lambda t}(\lambda t)^n g(\lambda)d\lambda}\end{aligned} \tag{1.173}$$

Moreover,

$$\begin{aligned}\Pr\{N(t)>n\} = \sum_{j=n+1}^\infty \Pr\{N(t)=j\} &= \\ =\sum_{j=n+1}^\infty \int_0^\infty e^{-\lambda t}\frac{(\lambda t)^j}{j!}g(\lambda)d\lambda &= \\ =\int_0^\infty \sum_{j=n+1}^\infty e^{-\lambda t}\frac{(\lambda t)^j}{j!}g(\lambda)d\lambda\end{aligned} \tag{1.174}$$

But $\Pr\{N(t) > n \mid \tilde{\lambda} = \lambda\} = \Pr\{\tau_{n+1} < t \mid \tilde{\lambda} = \lambda\}$ or

$$\sum_{j=n+1}^{\infty} e^{-\lambda t}\frac{(\lambda t)^j}{j!} = \int_0^t \lambda e^{-\lambda x}\frac{(\lambda x)^n}{n!}dx \quad (\star) \tag{1.175}$$

given that τ_{n+1} is Gamma$(n+1, \lambda)$ distributed and, being λ and t interchangeable

$$(\star) \quad = \int_0^\lambda te^{-tx}\frac{(tx)^n}{n!}dx$$

so that

$$\begin{aligned} \Pr\{N(t) > n\} &= \int_0^\infty \int_0^\lambda te^{-tx}\frac{(tx)^n}{n!}dxg(\lambda)d\lambda = \\ &= \text{(interchanging the order of integration)} = \\ &= \int_0^\infty te^{-t\lambda}\frac{(t\lambda)^n}{n!}(1 - G(\lambda))d\lambda \end{aligned} \tag{1.176}$$

Remark 1.9a. Stochastic intensity
Cox (1955) assumes that the intensity is not a r.v. but a stochastic process $\tilde{\lambda}(t)$. In this case (see Cox and Lewis (1966), p. 180)

$$\mathbb{E}[N(t)] = \mathbb{E}_\lambda\left[\mathbb{E}\left[N(t) \mid \lambda(u), 0 \le u \le t\right]\right] = \mathbb{E}_\lambda\left[\int_0^t \tilde{\lambda}(u)du\right] \tag{1.177}$$

Remark 1.9b. Hawkes processes; see § 1.9
A Hawkes process is a particular Poisson process $\tilde{N}(t)$ driven by a stochastic *intensity* function $\tilde{\lambda}(t)$ that depends, at any time $t > 0$, on the past history of the process itself. The occurrence at times $\tau_1 < ... < \tau_n < t$ of n Poisson events drives the process, increasing the intensity by a magnitude (typically positive) α, which is fading away between events, (typically) exponentially at a rate β, i.e.:

$$\begin{cases} \tilde{\lambda}(t) = \mu + \sum_{i=1}^{N(t)} \alpha_i e^{-\beta(t-\tau_i)}, & t = \tau_1, ..., \tau_{N(t)} \\ d\tilde{\lambda}(t) = \beta(\mu - \tilde{\lambda})dt, & \tau_{i-1} < t < \tau_i \end{cases} \tag{1.178}$$

We have

$$\mathbb{E}[N(t)] = \int_0^t \mathbb{E}[\tilde{\lambda}(s)]ds \tag{1.179}$$

1.5 Compound Poisson Processes (CPP)

This paragraph provides the basic theory of Compound Poisson processes and some applied examples.

1.5.1 Standard Compound Poisson Process

Let $N(t)$ be the number of Poisson events in a fixed time interval [0, t]. Let τ_j, $j = 1, 2, ..., N(t)$ be the arrival time of event j.

If $X_{\tau_j} \equiv X_j$ are i.i.d. r.v.'s independent from $N(t)$, we can define the compound Poisson process as

$$Z(t) = \sum_{j=1}^{N(t)} X_j, \qquad N(t) > 0 \tag{1.180}$$

As a typical example, $N(t)$ is the number of claims generated by an insurance policy in the interval $[0, t)$, and X_j is the claim amount related to the j-th accident, so that $Z(t)$ is the total amount of claims generated by the policy in the given time window $[0, t)$.

Analogously, if $N(t)$ is the number of customers arrived to a shop and X_j is the amount spent by customer j, then $Z(t)$ is the total expenditure made at the shop in $(0, t)$ by all customers.

Remark 1.10. When X_j are rewards (costs) received (paid) each time an event occurs, $Z(t)$ is the total reward earned (paid) by time t and it is called the Poisson Reward process.

Example. Customers arrive at a single-server bank according to a Poisson process with intensity λ. If the server is free, the customer enters the bank, otherwise they go home.

The time spent in the bank has distribution G with mean μ_G and the amount he or she deposits D_j has distribution H with mean μ_H. The accumulation of deposits is

$$Z(t) = \sum_{j=1}^{N(t)} D_j, \tag{1.181}$$

the average time between customers is

$$\mu = \mu_G + \frac{1}{\lambda}, \tag{1.182}$$

the average accumulation is

$$\mathbb{E}[Z(t)] = \lambda t \mu_H. \tag{1.183}$$

Example. In (1.180), if the X_j are Exponential r.v.'s (with rate parameter μ), then

$$\begin{aligned}\Pr\left\{Z(t)=\sum_{j=1}^{N(t)}X_j<z\right\}&=\mathbb{E}\left[\Pr\left\{\sum_{j=1}^{N(t)}X_j<z\,\middle|\,N(t)\right\}\right]=\\&=\sum_{k=0}^{\infty}\Pr\left\{\sum_{j=1}^{k}X_j<z\right\}\cdot\Pr\{N(t)=k\}=\\&=\sum_{k=0}^{\infty}\left(\int_0^z\frac{\mu^k}{\Gamma(k)}e^{-\mu x}x^{k-1}dx\right)\cdot e^{-\lambda t}\frac{(\lambda t)^k}{k!}\end{aligned}\tag{1.184}$$

and, hence,

$$\begin{aligned}\Pr\{Z(t)<z\}&=\int_0^z e^{-\mu x}e^{-\lambda t}\sum_{k=1}^{\infty}\frac{\mu^k x^{k-1}}{k!(k-1)!}(\lambda t)^k dx=\\&=e^{-\lambda t}\int_0^z e^{-\mu x}\sum_{k'=0}^{\infty}\frac{\mu^{k'+1}x^{k'}}{(k'+1)!k'!}(\lambda t)^{k'+1}dx=\\&=e^{-\lambda t}\int_0^z\frac{e^{-\mu x}}{x}\sum_{k=0}^{\infty}\frac{(\mu x\lambda t)^{k+1}}{k!(k+1)!}dx=\\&=e^{-\lambda t}\int_0^z\frac{e^{-\mu x}}{x}\sum_{k=0}^{\infty}\frac{\left(\sqrt{\mu x\lambda t}\right)^{2k+2}}{k!(k+1)!}dx=\\&=e^{-\lambda t}\int_0^z\frac{e^{-\mu x}}{x}\sum_{k=0}^{\infty}\frac{\sqrt{\mu x\lambda t}}{k!(k+1)!}\left(\frac{2\sqrt{\mu x\lambda t}}{2}\right)^{2k+1}dx=\\&=e^{-\lambda t}\int_0^z\frac{e^{-\mu x}}{x}\sqrt{\mu\lambda t x}\cdot I_1\left(2\sqrt{\mu\lambda t x}\right)dx\end{aligned}\tag{1.185}$$

where $I_1(x)=\sum_{k=0}^{\infty}\frac{\left(\frac{x}{2}\right)^{2k+1}}{k!(k+1)!}$ is the modified Bessel function of the first kind and here $x=2\sqrt{\lambda t\mu x'}$.

The following theorem provides the mean and variance of the Compound Poisson Process.

Theorem 1.13. *If X_j are i.i.d. and independent from $N(t)$ we have that*

$$\begin{cases} \mathbb{E}[Z(t)] = \mathbb{E}[N(t)] \cdot \mathbb{E}[X] \\ \mathbb{V}ar[Z(t)] = (\mathbb{E}[X])^2 \cdot \mathbb{V}ar[N(t)] + \mathbb{E}[N(t)] \cdot \mathbb{V}ar[X] \end{cases} \tag{1.186}$$

Proof. The first result is straightforward. The mean indeed reads

$$\begin{aligned} \mathbb{E}[Z(t)] = \mathbb{E}[\mathbb{E}[Z(t) \mid N(t)]] = \mathbb{E}\left[\mathbb{E}\left[\sum_{j=1}^{N} X_j \,\middle|\, N(t)\right]\right] = \\ = \mathbb{E}[N(t)\mathbb{E}[X \mid N(t)]] = \mathbb{E}[N(t)]\mathbb{E}[X] \end{aligned} \tag{1.187}$$

For the second result, we first need to evaluate the second moment

$$\begin{aligned} \mathbb{E}\left[Z^2(t)\right] = & \\ & = \mathbb{E}\left[\left(\sum_{h=1}^{N(t)} X_h\right)^2\right] = \\ & = \mathbb{E}\left[\sum_{h=1}^{N(t)} X_h^2 + \sum_{i\neq j}^{N(t)} X_i X_j\right] = \\ & = \mathbb{E}\left[X^2\right] \cdot \mathbb{E}[N(t)] + \mathbb{E}\left[N(t)(N(t)-1) \cdot X_i X_j\right] = \\ & = \mathbb{E}[X^2]\mathbb{E}[N(t)] + \left(\mathbb{E}[N^2(t)] - \mathbb{E}[N(t)]\right)\left(\mathbb{E}[X]\right)^2 \end{aligned} \tag{1.188}$$

Thus

$$\begin{aligned} \mathbb{V}ar\left[Z(t)\right] = & \\ & = \mathbb{E}[Z^2(t)] - (\mathbb{E}Z(t))^2 = \\ & = \mathbb{E}[X^2]\mathbb{E}[N(t)] + (\mathbb{E}[X])^2\left(\mathbb{E}[N^2(t)] - \mathbb{E}[N(t)]\right) + \\ & - \left(\mathbb{E}[N(t)]\right)^2\left(\mathbb{E}[X]\right)^2 = \\ & = \mathbb{E}[N(t)]\left(\mathbb{E}[X]^2 - (\mathbb{E}[X])^2\right) + \\ & + (\mathbb{E}[X])^2\left(\mathbb{E}[N^2(t)] - (\mathbb{E}[N(t)])^2\right) = \\ & = \mathbb{E}[N(t)]\mathbb{V}ar[X] + \mathbb{E}[X]^2\mathbb{V}ar[N] \end{aligned} \tag{1.189}$$

□

Remark 1.11.
If $N(t)$ is a renewal process the theorem holds asymptotically (Ross, 2024, p. 459).

The process

$$M(t) = Z(t) - \lambda t \mathbb{E}[X] \tag{1.190}$$

is referred to as a *compensated* compound Poisson process, and it is a zero-mean martingale, given that

$$\mathbb{E}\left[M(t) \mid M(s)\right] = M(s) \qquad \forall t \geq s \tag{1.191}$$

An alternative derivation is provided by the following theorem.

Theorem 1.14. *The mean and variance of $Z(t)$ are:*

$$\begin{aligned} \mathbb{E}[Z(t)] &= \mathbb{E}[N(t)]\,\mathbb{E}[X] \\ \mathbb{V}ar[Z(t)] &= (\mathbb{E}[X])^2 \mathbb{V}ar[N(t)] + \mathbb{E}[N(t)] \mathbb{V}ar[X] \end{aligned} \tag{1.192}$$

Proof. For the mean, see Theorem 1.13. □

For the variance, we use the following

Lemma 1.1. *(Law of total variance):*

$$\begin{aligned} \mathbb{V}ar[Z(t)] &= \mathbb{V}ar[X_1 + X_2 + ... + X_{N(t)}] = \\ &= (\mathbb{E}[X])^2 \mathbb{V}ar[N(t)] + Var[X]\mathbb{E}[N(t)] \end{aligned} \tag{1.193}$$

Proof. For any $t > 0$ and $N = N(t) > 0$ a homogeneous Poisson process, the expression obtains after arranging the first and second moments

$$\begin{cases} \mathbb{E}[X_1 + ... + X_N] = \mathbb{E}[X]\mathbb{E}[N] \\ \mathbb{E}[(X_1 + ... + X_N)^2] = \mathbb{E}[\mathbb{E}[(X_1 + ... + X_N)^2] \mid N] = \\ = \mathbb{E}[N \mathbb{E} X^2 + N(N-1)(\mathbb{E}[X])^2] \end{cases} \tag{1.194}$$

□

Lemma 1.2. *(Conditional variance formula). For arbitrary r.v.'s Z, Y the following decomposition holds,*

$$\mathbb{V}ar[Z] = \mathbb{E}_Y[\mathbb{V}ar[Z \mid Y]] + \mathbb{V}ar_Y[\mathbb{E}[Z \mid Y]]. \tag{1.195}$$

Proof. The expression is straightforward:

$$\begin{aligned} \mathbb{V}ar[Z] = \mathbb{E}[(Z - \mathbb{E}[Z])^2] = \\ = \mathbb{E}[(Z - \mathbb{E}[Z \mid Y] + \mathbb{E}[Z \mid Y] - \mathbb{E}[Z])^2] = \\ = \mathbb{E}_Y[\mathbb{E}[(Z - \mathbb{E}[Z \mid Y])^2 + (\mathbb{E}[Z \mid Y] - \mathbb{E}[Z])^2 \mid Y] + \\ +2 \cdot \mathbb{E}_Y[E[(Z - E[Z \mid Y]) \cdot (E[Z \mid Y] - E[Z]) \mid Y]] \end{aligned} \tag{1.196}$$

but the second component is zero being

$$2(\mathbb{E}[Z \mid Y] - \mathbb{E}[Z]) \cdot \mathbb{E}[Z - E[Z \mid Y] \mid Y] = 0. \tag{1.197}$$

Moreover,

$$\mathbb{E}_Y[\mathbb{E}[(Z - \mathbb{E}[Z \mid Y])^2 \mid Y]] = \mathbb{E}_Y[\mathbb{V}ar[Z \mid Y]] \tag{1.198}$$

and

$$\begin{aligned} \mathbb{E}_Y[\mathbb{E}[(\mathbb{E}[Z \mid Y] - \mathbb{E}[Z])^2 \mid Y]] = \mathbb{E}[(\mathbb{E}[Z \mid Y] - \mathbb{E}[Z])^2] = \\ = \mathbb{V}ar_Y[\mathbb{E}[Z \mid Y]] \end{aligned} \tag{1.199}$$

so that $\mathbb{V}ar[Z] = \mathbb{V}ar_Y[\mathbb{E}[Z \mid Y]] + \mathbb{E}_Y[\mathbb{V}ar[Z \mid Y]]$.

The variance is decomposed in the variance of the conditional mean plus the mean of the conditional variance.

Applying Lemma 1 to Theorem 1.5

$$\mathbb{V}ar[Z(t)] = \mathbb{V}ar_N[\mathbb{E}[Z(t) \mid N(t)]] + \mathbb{E}_N[\mathbb{V}ar[Z(t) \mid N(t)]] \tag{1.200}$$

so that

$$\begin{cases} \mathbb{V}ar_N[\mathbb{E}[Z(t) \mid N(t)]] = \mathbb{V}ar_N[N(t)\mathbb{E}[X]] = (\mathbb{E}[X])^2 \mathbb{V}ar[N(t)], \\ \mathbb{E}_N[\mathbb{V}ar[Z(t) \mid N(t)]] = \mathbb{E}_N[N(t)\mathbb{V}ar[X]] = \mathbb{E}[N(t)]\mathbb{V}ar[X]. \end{cases} \tag{1.201}$$

We used the property $\mathbb{E}_Y[\mathbb{E}[Z \mid Y]] = \mathbb{E}[Z]$ of the conditional expectation.

In particular, if $Z = 1_A$ is the indicator function of event A,

$$\mathbb{E}_Y[Pr\{A \mid Y\}] = Pr\{A\} = \mathbb{E}[Z] \tag{1.202}$$

□

Theorem 1.15. *If $N(t)$ is a Poisson r.v. and the X_j are non-negative i.i.d. r.v.'s with probability distribution function F, and $N(t)$ is independent of the X's, we have that the probability generating function of the compound Poisson process is:*

$$\begin{aligned}\mathbb{E}\left[u^{Z(t)}\right] &= \exp\left\{-\lambda t + \lambda t \mathbb{E}\left[u^X\right]\right\} \\ &= \exp\left\{-\lambda t\int_0^\infty (1-u^x)dF(x)\right\}.\end{aligned} \tag{1.203}$$

Proof. In fact

$$\begin{aligned}\mathbb{E}\left[u^{Z(t)}\right] = \mathbb{E}\left[u^{\sum_{j=1}^{N(t)} X_j}\right] = \\ = \mathbb{E}_N\left[\mathbb{E}\left[u^{\sum_{j=1}^{N(t)} X_j}|N(t)\right]\right] = \mathbb{E}_N\left[\mathbb{E}\left[u^X\right]^{N(t)}\right] = \\ = \sum_{k=0}^{\infty}\left(\mathbb{E}\left[u^X\right]\right)^k \cdot \Pr\{N(t)=k\} = \sum_{k=0}^{\infty}\mathbb{E}\left[u^X\right]^k \cdot e^{-\lambda t}\frac{(\lambda t)^k}{k!} = \\ = e^{-\lambda t}e^{\lambda t\cdot\mathbb{E}[u^X]} = e^{-\lambda t}e^{\lambda t\cdot\int_0^\infty u^x dF(x)} = \\ = e^{-\lambda t\left(1-\int_0^\infty u^x dF(x)\right)} = e^{-\lambda t\int_0^\infty (1-u^x)dF(x)}.\end{aligned} \tag{1.204}$$

and the result obtains. □

Remark 1.12. (see §1.8).
If $X_j = N_j(t)$ the process $Z(t) = \sum_{j=1}^{N(t)} N_j(t)$ is called compound Poisson-Poisson. It represents the Poisson sum of Poisson processes.

If X_j is finite or countable with

$$\Pr\{X_j = a_h\} = p_h, \qquad \sum_{h\geq 1} p_h = 1, \qquad \forall j = 1,2,... \tag{1.205}$$

and $N_h(t)$ is the number of a_h-results, then $N_h(t)$ are independent Poisson processes with intensity λp_h and the compound Poisson can be represented as:

$$Z(t) = \sum_{h\geq 1} a_h N_h(t) \tag{1.206}$$

Remark 1.12. (cont.)
Note that, in fact:

$$\begin{cases}\mathbb{E}[Z(t)] = \sum_{h\geq 1} a_h \mathbb{E}[N_h(t)] = \sum_{h\geq 1} a_h \lambda p_h t = \lambda t \mathbb{E}[X] \\ \mathbb{V}ar[Z(t)] = \sum_{h\geq 1} a_h^2 \mathbb{V}ar[N_h(t)] = \sum_{h\geq 1} a_h^2 \lambda p_h t = \lambda t \mathbb{E}[X^2]\end{cases} \tag{1.207}$$

In the special case of Poisson(θ) distributed r.v.'s $X_j = N_j$, $\forall j$, then

$$\begin{cases}\mathbb{E}[Z(t)] = \lambda t\theta \\ \mathbb{V}ar[Z(t)] = \lambda t\theta(1+\theta) \quad (overdispersed)\end{cases} \tag{1.208}$$

and $\Pr\{Z(t) = n\} = \frac{e^{-\lambda t}\theta^n}{n!} \sum_{j=0}^{\infty} \frac{(\lambda t e^{-\theta})^j j^n}{j!} \qquad n = 0, 1, ...$
which is the Neyman type A distribution (see Theorem 1.19).

Remark 1.13. Compound logarithmic distribution
If $N(t)$ is Poisson(λt) and X_j are i.i.d. logarithmic distributed r.v.'s $\forall j$, i.e.

$$\Pr\{X_j = k\} = \frac{-p^k}{k \ln(1-p)} \tag{1.209}$$

then

$$Z(t) = \sum_{j=1}^{N(t)} X_j \sim NegBin(r, p) \tag{1.210}$$

with $r = \frac{\lambda t}{-\ln(1-p)}$.

Theorem 1.16. *If $Z_1(t)$ and $Z_2(t)$ are two independent compound Poisson processes with intensity λ_i and distribution F_{Z_i}, $i = 1, 2$, then $Z(t) = Z_1(t) + Z_2(t)$ is also a compound Poisson process with* *intensity* $\lambda_1 + \lambda_2$ *and distribution*

$$F_Z(x) = \frac{\lambda_1}{\lambda_1 + \lambda_2} F_{Z_1}(x) + \frac{\lambda_2}{\lambda_1 + \lambda_2} F_{Z_2}(x) \tag{1.211}$$

Remark 1.14.
For a portfolio of Q policies the total aggregate claim is $Z(t) = \sum_{q=1}^{Q} \sum_{j=1}^{N_q(t)} X_{q,j}$, where the $X_{q,j}$'s are i.i.d. Hence mean, variance and p.g.f. are given respectively as

$$\begin{aligned} \mathbb{E}[Z(t)] &= \textstyle\sum_q \mathbb{E}[N_q(t)]\mathbb{E}[X_q], \\ \mathbb{V}\mathrm{ar}[Z(t)] &= \textstyle\sum_q \mathbb{V}\mathrm{ar}[N_q(t)] \cdot [(\mathbb{E}[X_q])^2 + \mathbb{V}\mathrm{ar}(X_q)], \\ \mathbb{E}\left[u^{\sum_q Z_q(t)}\right] &= e^{-t\sum_q \lambda_q (1-\mathbb{E}u^{X_q})} = e^{-t\sum_q \lambda_q \int_0^\infty (1-u^x)dF_q(x)}. \end{aligned}$$

Multiplicative Compound Poisson. Let $S(t)$ be the price of an asset at time t subject to random shocks according to the multiplicative model

$$S(t) = S(0) \prod_{i=1}^{N(t)} X_i \tag{1.212}$$

where $N(t)$ is a Poisson process with intensity λ and X_i are i.i.d. Exponential r.v.'s with parameter μ.

Then

$$\begin{aligned} \mathbb{E}[S(t) \mid N(t) = n] &= S(0)\mathbb{E}\left[\prod_{i=1}^{N(t)} X_i \mid N(t) = n\right] = \\ &= S(0)\mathbb{E}\left[\prod_{i=1}^{n} X_i\right] = S(0)\mathbb{E}[X]^n = S(0)\left(\frac{1}{\mu}\right)^n \end{aligned} \tag{1.213}$$

and

$$\begin{aligned} \mathbb{E}[S(t)] &= S(0) \sum_{n=1}^{\infty} \left(\frac{1}{\mu}\right)^n e^{-\lambda t} \frac{(\lambda t)^n}{n!} = \\ &= S(0)e^{-\lambda t} \sum_{n=1}^{\infty} \frac{\left(\frac{\lambda t}{\mu}\right)^n}{n!} = \\ &= S(0)e^{-\lambda t} e^{\frac{\lambda t}{\mu}} \end{aligned} \tag{1.214}$$

Moreover,

$$\begin{cases} \mathbb{E}[S^2(t) \mid N(t) = n] = S^2(0)(\mathbb{E}[X^2])^n = S^2(0)\left(\frac{2}{\mu^2}\right)^n \\ \mathbb{E}[S^2(t)] = S^2(0)e^{-\lambda t}e^{\frac{2\lambda t}{\mu^2}} \end{cases} \tag{1.215}$$

1.5.2 Distribution of the Maximum

If to each Poisson event (e.g. a car accident) we associate a r.v. X_j (the amount of the damage), it is of interest to calculate the following probability:

$$\Pr\left\{\max\left(X_1, \ldots, X_{N(t)}\right) < b\right\} \tag{1.216}$$

which represents the maximal damage produced by the $N(t)$ events in the interval $(0, t)$.

Note that for $N(t) = 0$ we assume that $\max\left(X_1, \ldots, X_{N(t)}\right) = 0$.

The probability (1.216) can be calculated as

$$\begin{aligned} \Pr\left\{\max\left(X_1, \ldots, X_{N(t)}\right) < b\right\} &= \\ = \mathbb{E}\left[\Pr\left\{\max\left(X_1, \ldots, X_{N(t)}\right) < b \mid N(t)\right\}\right] &= \\ = \sum_{k=0}^{\infty} \Pr\left\{\max\left(X_1, \ldots, X_k\right) < b\right\} \Pr\{N(t) = k\} &= \\ = \sum_{k=0}^{\infty} [\Pr(X < b)]^k e^{-\lambda t}\frac{(\lambda t)^k}{k!} &= \\ = e^{-\lambda t + \lambda t \Pr\{X<b\}} = e^{-\lambda t \cdot Prob\{X>b\}}&. \end{aligned} \tag{1.217}$$

Note that the distribution function of the random variable $\max\left(X_1, \ldots, X_{N(t)}\right)$ reads

$$\Pr\left\{\max\left(X_1, \ldots, X_{N(t)}\right) < b\right\} = \begin{cases} 0 & b < 0 \\ e^{-\lambda t} & b = 0 \\ e^{-\lambda t + \lambda t \Pr(X<b)} & b > 0 \end{cases} \tag{1.218}$$

and shows a jump of height $e^{-\lambda t}$ at $b = 0$.

In the special case where X is exponentially distributed, that is

$$\Pr(X < b) = \mu \int_0^b e^{-\mu y} dy = 1 - e^{-\mu b}, \tag{1.219}$$

for $b > 0$, the probability distribution (1.218) becomes

$$\Pr\left\{\max\left(X_1, \ldots, X_{N(t)}\right) < b\right\} = \begin{cases} 0 & b < 0 \\ e^{-\lambda t} & b = 0 \\ e^{-\lambda t e^{-\mu b}} & b > 0 \end{cases} \tag{1.220}$$

and is known as the Gumbel or Gnedenko distribution.

Theorem 1.17. *Let $N_{\alpha,\delta}$ be a Poisson r.v. with mean value (i.e. rate parameter) $\mathbb{E}[N_{\alpha,\delta}] = \delta^{-\alpha}$, $0 < \alpha < 1$. Let $X_{\delta,k}$, $k = 1, 2, ...,$ be a sequence of independent r.v.'s with*

$$Pr\{X_{\delta,k} > x\} = \begin{cases} 1 & 0 < x \leq \delta \\ \frac{\delta^\alpha}{x^\alpha} & x > \delta \end{cases} \tag{1.221}$$

with probability density $Pr\{X_{1,k} \in dx\}/dx = \alpha x^{-(\alpha+1)}\delta^\alpha$.

Then the sequence of random sums

$$Z_{\alpha,\delta} = \sum_{k=1}^{N_{\delta,\alpha}} X_{\delta,k} \tag{1.222}$$

converges in distribution, as $\delta \to 0$, to the Stable *r.v. (subordinator) $Z \sim S_\alpha(\sigma, 0, 1)$, where $\sigma = \left(\Gamma(1-\alpha)\cos\left(\frac{\pi\alpha}{2}\right)\right)^{\frac{1}{\alpha}}$.*

Proof. The *characteristic function* of the Z r.v. is obtained as follows:

$$\begin{aligned} \mathbb{E}\left[e^{iZ_{\delta,\alpha}t}\right] &= \mathbb{E}\left[\mathbb{E}\left(e^{i\sum_{k=1}^{N_{\delta,\alpha}} X_{\delta,k}t}\right) \,\middle|\, N_{\delta,\alpha}\right] = \\ = \mathbb{E}\left[\left(\mathbb{E}e^{iX_\delta t}\right)^{N_{\delta,\alpha}}\right] &= \sum_{h=0}^{\infty} \mathbb{E}\left(e^{iX_\delta t}\right)^h e^{-\delta^{-\alpha}} \frac{(\delta^{-\alpha})^h}{h!} = \\ &= \exp\left(-\delta^{-\alpha} + \delta^{-\alpha}\mathbb{E}\left(e^{iX_\delta t}\right)\right) = \\ &= \exp\left(\alpha \int_\delta^\infty x^{-(\alpha+1)}(e^{itx} - 1)dx\right) \end{aligned} \tag{1.223}$$

whence the limit for $\delta \to 0$ of the *characteristic function* of $Z_{\delta,\alpha}$ is

$$\lim_{\delta\to 0} \mathbb{E}\left[e^{iZ_{\delta,\alpha}t}\right] = \exp\left(\alpha \int_0^\infty x^{-(\alpha+1)}(e^{itx}-1)dx\right) \tag{1.224}$$

Clearly the above (improper) integral can be regarded as

$$\lim_{u\to 0} \int_0^\infty x^{-(\alpha+1)}(e^{-ux+itx}-1)dx; \tag{1.225}$$

moreover, the above integral reduces to

$$\begin{aligned} &\int_0^\infty x^{-(\alpha+1)}(e^{-(u-it)x})dx = \\ &= -(u-it)\int_0^\infty \left(\int_0^x e^{-y(u-it)}dy\right) x^{-(\alpha+1)}dx = \\ &= -(u-it)\int_0^\infty e^{-y(u-it)}\left(\int_y^\infty x^{-\alpha-1}dx\right)dy = \\ &= -\frac{(u-it)}{\alpha}\int_0^\infty e^{-y(u-it)}y^{-\alpha}dy = \\ &= (\text{for } y(u-it)=w) = \\ &= -\frac{(u-it)^\alpha}{\alpha}\int_0^\infty e^{-w}w^{(1-\alpha)-1}dw = \\ &= -\frac{\Gamma(1-\alpha)}{\alpha}(u-it)^\alpha. \end{aligned} \tag{1.226}$$

Since the polar transformations hold

$$u - it = \begin{cases} \sqrt{u^2+t^2}e^{i\arctan\left(-\frac{|t|}{u}\right)}, & t>0 \\ \sqrt{u^2+t^2}e^{i\arctan\left(\frac{|t|}{u}\right)}, & t<0 \end{cases} \tag{1.227}$$

then

$$\lim_{u\to 0}(u-it)^\alpha = \begin{cases} |t|^\alpha e^{-i\frac{\pi\alpha}{2}}, & t>0 \\ |t|^\alpha e^{i\frac{\pi\alpha}{2}}, & t<0 \end{cases} \tag{1.228}$$

Putting together the above results, the expression becomes

$$\begin{aligned}\mathbb{E}\left[e^{itZ_{\delta,\alpha}}\right] &= \exp\left(-\alpha\frac{\Gamma(1-\alpha)}{\alpha}\cdot\mid t\mid^{\alpha}\right)\cdot\exp\left(-i\frac{\pi}{2}\alpha\cdot\operatorname{sgn}(t)\right) = \\ &= \exp\left(-|t|^{\alpha}\Gamma(1-\alpha)\left[\cos\left(\frac{\pi}{2}\alpha\right) - i\operatorname{sgn}(t)\sin\left(\frac{\pi}{2}\alpha\right)\right]\right) = \\ &= \exp\left(-\mid t\mid^{\alpha}\Gamma(1-\alpha)\right)\cdot\exp\left(-i\frac{\pi}{2}\alpha\operatorname{sgn}(t)\right)\end{aligned} \tag{1.229}$$

that coincides with the *characteristic function* of the $(\sigma, 1, 0)$ α-Stable r.v. for $\sigma^{\alpha} = \Gamma(1-\alpha)\cos\left(\frac{\pi}{2}\alpha\right)$, $\mu = 0$ and $\beta = 1$.

This completes the proof of the convergence of the compound random sum $Z_{\delta,\alpha}$ to the α-Stable r.v. $Z \sim S_{\alpha}(\sigma, 1, 0)$.

□

Remark 1.15. Analogously, for the Laplace transform of Z

$$\mathbb{E}\left[e^{-\theta Z_{\delta,\alpha}}\right] = \exp\left(\alpha\int_{\delta}^{\infty} x^{-(\alpha+1)}(e^{-\theta x} - 1)dx\right) \tag{1.230}$$

it holds true that the limit for $\delta \to 0$ is

$$\begin{aligned}\mathbb{E}\left[e^{-\theta Z}\right] &= \exp\left(\alpha\int_{0}^{\infty}(e^{-\theta x} - 1)x^{-(\alpha+1)}dx\right) = \\ &= \exp\left(-\theta\left(\int_{0}^{\infty} e^{-\theta x}x^{-\alpha}dx\right)\right) = \\ &= \exp\left(-\left(\int_{0}^{\infty} e^{-y}y^{-\alpha}dy\right)\theta^{\alpha}\right) = \\ &= \exp\left(-\Gamma(1-\alpha)\theta^{\alpha}\right) = \exp\left(-\frac{\sigma^{\alpha}}{\cos\left(\frac{\pi\alpha}{2}\right)}\theta^{\alpha}\right)\end{aligned} \tag{1.231}$$

1.5.3 Classical Ruin Theory

The *classical ruin theory* is a mathematical construct formulated by Scandinavian mathematicians (Cramér, Lundberg) in the beginning of the XX century. Its main risk process is the *risk reserve process* $R(t)$ defined as

$$R(t) = u + \pi t - \sum_{i=1}^{N(t)} X_i, \qquad t \geq 0 \tag{1.232}$$

where u is the initial capital of the insurance company, π denotes the *premium intensity* (premiums per unit time) and the compound sum denotes the *aggregate claim amount* up to time t.

In the Cramér (1930)-Lundberg (1903) formulation the aggregate claim process is made up of a *Poisson* process $N(t)$ with parameter λ and a *claim size* r.v.'s X_i with common distribution $F_X(x)$ with mean μ_X, so that $\sum_{i=1}^{N(t)} X_i$ is a Compound Poisson process (CPP).

One of the main objectives of the theory is to compute the ruin probability ψ as a function of the initial capital u, defined as, in infinite horizon:

$$\psi(u) = \Pr\left\{\inf_{t\geq 0} R_t < 0\right\} \tag{1.233}$$

and, in finite time, as:

$$\psi(u, T) = \Pr\left\{\inf_{0\leq t\leq T} R_t < 0\right\} \tag{1.234}$$

When $R(t) < 0$ the insurer is ruined (insolvent) and a necessary condition for $\psi(u) < 1$ is $\pi > \lambda\mu_X$ (*net profit condition*).

Considering the time to ruin (as a function of capital u)

$$\tau(u) = \inf\{t > 0 : R_t < 0 \mid R_0 = u\} \tag{1.235}$$

then $\psi(u)$ and $\psi(u, T)$ are also equal respectively to $\Pr\{\tau(u) < \infty\}$ and $\Pr\{\tau(u) \leq T\}$.

The *average claim amount per unit time* is:

$$\frac{1}{t}\mathbb{E}\left[\sum_{i=1}^{N(t)} X_i\right] = \lambda\mu_X = \rho \tag{1.236}$$

(see (1.186)) and it is related to the (relative) *safety loading* θ, i.e. $\theta = \frac{\pi-\rho}{\rho}$.

In the literature, different point processes have been applied as arrival processes: apart from the classical Poisson process, a Cox process is the typical alternative (see Asmussen and Albrecher (2010), Example 1.2); recently, the ruin model driven by the *time-fractional Poisson process* has been studied, yielding to analytical solutions for the ruin probability (see Chapter 5 for an application to earthquakes data). The Poisson process is discussed in the present section.

In the case of the ruin model driven by the Poisson process, the moments are more readily expressed in terms of the *claims surplus process* $S_t = u - R_t$ and are related to the (non-central) moments $\mu_X^{(k)}$ of the claim size distribution (Albrecher and Asmussen, 2010) as follows (note the linearity of the moments in time):

$$\begin{cases} \mathbb{E}[S_t] = (\lambda\mu_X - 1)\cdot t = (\rho - 1)\cdot t \\ \mathbb{V}ar[S_t] = \lambda\mu_X^{(2)}\cdot t \end{cases} \tag{1.237}$$

The so called *Cramér-Lundberg approximation* of the ultimate ruin probability $\psi(u)$ reads

$$\psi(u) \sim Ce^{-\mathcal{R}u} \qquad u\to\infty \tag{1.238}$$

where C and $\mathcal{R}$ are coefficients related to the m.g.f. $\hat{F}(\gamma)$ of the claim size distribution. In particular, $\mathcal{R}$ is called *adjustment coefficient* and arises as the unique solution to the *Lundberg equation*

$$\mathcal{R}: \begin{cases} \hat{F}(\mathcal{R}) = 1 + \frac{\mathcal{R}}{\lambda} \qquad \text{or} \\ \lambda(\hat{F}(\mathcal{R}) - 1) - \mathcal{R} = 0 \end{cases} \tag{1.239}$$

and it is also referred to as the *Lundberg exponent* (see Albrecher and Asmussen, 2010).

Bounds of the ruin probability are less precise (and easier to obtain). The main result is the celebrated *Lundberg's inequality*

$$\psi(u) \leq e^{-\mathcal{R}u} \tag{1.240}$$

Its precision is high not only for large values of the initial capital u, but for all $u > 0$ (see Albrecher and Asmussen (2010)).

Finally, we sketch a heuristic treatment of the *Pollaczeck-Khintchin* formula. Recalling that the time-to-ruin $\tau(u)$, as a function of initial capital u, can be defined in terms of the surplus process S_t (defined above) as $\tau(u) = \inf\{t > 0 : S_t > u\}$, define $M = \max_{t \geq 0} S_t$ so that $\psi(u) = \Pr\{M > u\}$.

The *Pollaczeck-Khintchin formula* provides a representation result for the ruin probability in terms of a geometric series of convolution powers (see Albrecher and Asmussen, 2010), i.e.[4]

$$\psi(u) = \Pr\{M > u\} = \left(1 - \frac{\rho}{\pi}\right) \sum_{n=1}^{\infty} \left(\frac{\rho}{\pi}\right)^n \bar{F}_{\mu}^{\star n}(u) \tag{1.241}$$

where $\bar{F}(x) = Pr\{X_i > x\}$ is the tail of the common *claim size distribution* $F_X(x)$ and $\bar{F}_{\mu}(x)$ stands for the *integrated tail distribution*, that is $\bar{F}_{\mu}(u) = \frac{\int_0^u \bar{F}(x)dx}{\mu_X}$ (see Asmussen and Albrecher (2010) for further details.

The Compound Poisson model with Exponential claim size distribution $F_X(x) = 1 - e^{-\frac{x}{\mu_X}}$ of mean μ_X brings to a closed form expression of the (*ultimate*) ruin probability, i.e.

$$\psi(u) = \frac{\lambda \mu_X}{\pi} \exp\left\{-\left(\frac{1}{\mu_X} - \frac{\lambda}{\pi}\right) u\right\} \tag{1.242}$$

[4]The *Pollaczeck-Khintchin* formula is a solution to a particular integral equation (see Rolski et Al., 1999, p. 164).

In this special case, for a given (ultimate) ruin probability ψ_0, the *minimum capital requirement* is

$$u \geq u_0 = \frac{\ln\left(\frac{\lambda\mu_X}{\pi}\right) - \ln(\psi_0)}{\frac{1}{\mu_X} - \frac{\lambda}{\pi}} \tag{1.243}$$

Remark 1.16. Convolution
The distribution of the sum of two independent r.v.'s with distribution F and G respectively is given by

$$(F \star G)(x) = \int_{-\infty}^{\infty} F(x-u)dG(u) \qquad \forall x \in \mathbb{R} \tag{1.244}$$

The density is

$$(f \star g)(x) = \int_{-\infty}^{\infty} f(x-u)g(u)du \qquad \forall x \in \mathbb{R} \tag{1.245}$$

The operator $F \star G$ is called convolution. In case of discrete probabilities $\{p_k, k \in \mathbb{N}\}$, $\{q_k, k \in \mathbb{N}\}$ we have

$$(p \star q)_k = \sum_{i+j=k} p_i q_j \qquad \forall k \in \mathbb{N} \tag{1.246}$$

The n-fold convolution of F is defined iteratively:

$$\begin{cases} F^{\star 0}(x) = \delta_0(x) \qquad \text{with } \delta_0(x) = \begin{cases} 1 & x \geq 0 \\ 0 & x < 0 \end{cases} \\ F^{\star n}(x) = F^{\star(n-1)}(x) \star F(x) \end{cases} \tag{1.247}$$

1.5.4 Estimation of a Compound Poisson Process Model for Fire Insurance

The Danish fire loss data provide the daily closing value for the Danish fire losses measured from January 1980 through December

1990 in millions of Danish Krones. The following table 1.7 resumes the occurrence of fires in the data:

Days with k fires, 1980-1990; $n = 4,018$		
k	n_k	np_k
0	2,373	2,344.2
1	1,219	1,263.1
2	343	340.3
3	72	61.1
4	11	9.2
–	$\lambda = 0.53$	$\chi_3^2 = 4.2$
–	$s^2 = 0.56$	p-value $= 0.24$

As for the loss amounts, the following table 1.8 reports the parameter estimation for a Lognormal, a Gamma and a Weibull distribution:

Parameter estimation on loss data		
LN(μ, σ^2)	Gamma(λ, ν)	Weibull(β, c)
$\mu = 0.79$ (0.015)	$\lambda = 0.383$ (0.013)	$\beta = 0.958$ (0.012)
$\sigma = 0.72$ (0.011)	$\nu = 1.297$ (0.035)	c $= 3.29$ (0.078)

As for the loss sizes, natural candidate loss distributions in risk theory are the Gamma (or Exponential), Lognormal and Weibull distributions. Preliminary data analysis shows that a Weibull distribution does not fit properly nor the extremes nor the center of the distribution, as depicted respectively by QQ- and PP-plots (see Fig. 1.10).

Considering the Lognormal distribution for the losses, parameter estimates for the resulting Compound Poisson model are shown in the table:

Danish fire losses: Compound Poisson-LogNormal model		
method	mean (S.E.)	std dev (S.E.)
MLE	0.7 (mln Danish Krones)	0.015
MME	0.7 (mln Danish Krones)	0.01

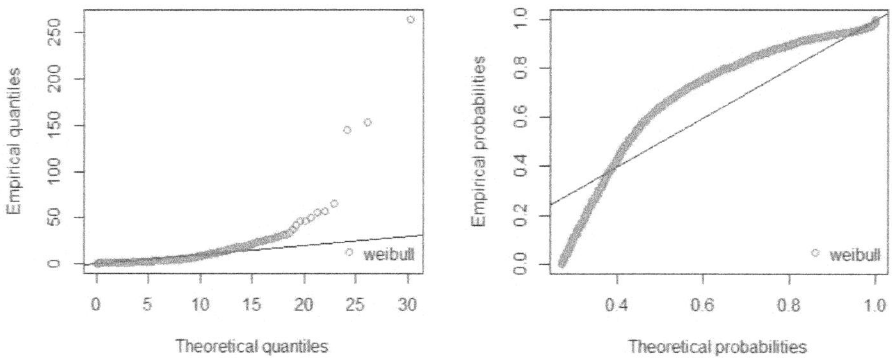

FIGURE 1.10
QQ and PP plots for the goodness-of-fit of a Weibull distribution to loss data.

Fig. 1.11 shows how not only the choice of the distribution matters (whether heavy-tailed or light-tailed) but a more subtle aspect related to the estimation method: here, the distribution function computed with MM estimators fits worse the left tail of the distribution (small claim sizes) compared to the distribution function obtained from ML estimators.

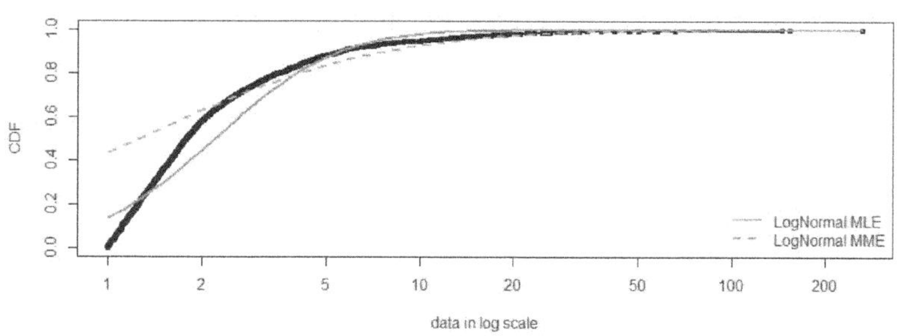

FIGURE 1.11
Empirical and fitted Compound Poisson-LogNormal CDF for the Danish fire loss (log-)data.

Note that due to large (*attritional*) claims, the Compound Poisson-LogNormal model does not fully capture the tail[5].

1.6 Poisson Random Fields

Points randomly distributed in the Euclidean space $\mathbb{R}^n$ can be modelled by means of a multi-index Poisson process, called Poisson field.

The basic property of the Poisson random fields (see Fig. 1.15) is represented by the independency of the random variables $N(S_A)$ and $N(S_B)$ for arbitrary disjoint sets S_A and S_B (that is $S_A \cap S_B = \emptyset$). We have that for the Poisson random field of rate $\lambda > 0$ we have that

$$\Pr\{N(S) = k\} = \frac{\lambda^k [\mu(S)]^k}{k!} e^{-\lambda\mu(S)} \qquad k \geq 0 \tag{1.248}$$

where $\mu(S)$ is the volume of the set $S \subset \mathbb{R}^n$ and $N : \mathcal{B}^n \to \mathbb{N} \cup \{0\}$ when $\mathcal{B}^n$ is the Borel class over $\mathbb{R}^n$.

Example. For a three-dimensional Poisson random field the distribution (1.248) becomes

$$\Pr\{N(S_r^3) = k\} = \exp\{-\lambda\mu(S_r^3)\} \frac{\lambda^k \left[\mu(S_r^3)\right]^k}{k!} \tag{1.249}$$

where $S_r^3 \equiv \{\underline{\mathbf{x}} \equiv \{x_1, x_2, ..., x_n\} :\| \underline{\mathbf{x}} \|_2^2 \leq r^2\}$.

The area of a spherical surface of $\mathcal{R}^n$ is

$$\text{Area}\,(S_r^n) = \frac{2\pi^{\frac{n}{2}}}{\Gamma(\frac{n}{2})} r^{n-1} \tag{1.250}$$

[5]The literature on Extreme Value Theory (EVT) has investigated several distributions in relation to the Danish fire losses data set and Pareto distributions (or other heavy-tailed distributions) fit the data properly.

This can be proved probabilistically by considering the following integral

$$\int_{\mathcal{R}^n}\int\cdots\int \frac{e^{-\frac{x_1^2}{2}}}{\sqrt{2\pi}}\frac{e^{-\frac{x_2^2}{2}}}{\sqrt{2\pi}}\cdots\frac{e^{-\frac{x_n^2}{2}}}{\sqrt{2\pi}}dx_1dx_2...dx_n = 1 \tag{1.251}$$

By passing to the hyperspherical coordinates the (1.251) integral becomes

$$\int_0^\infty \text{Area}(S_1)r^{n-1}\frac{e^{-\frac{r^2}{2}}}{(2n)^{\frac{n}{2}}}dr = 1 \tag{1.252}$$

where

$$\text{Area}(S_1) = \frac{(2\pi)^{\frac{n}{2}}}{\int_0^\infty e^{\frac{-r^2}{2}}r^{n-1}dr} = \frac{2^{\frac{n}{2}}\pi^{\frac{n}{2}}}{2^{\frac{n}{2}-1}\Gamma(\frac{n}{2})} = \frac{2\pi^{\frac{n}{2}}}{\Gamma(\frac{n}{2})} \tag{1.253}$$

Of course

$$\text{Area}(S_r^n) = \frac{2\pi^{\frac{n}{2}}r^{n-1}}{\Gamma(\frac{n}{2})} \tag{1.254}$$

and

$$\mathbb{V}ol[S_r^n] = \int_0^r \frac{2\pi^{\frac{n}{2}}}{\Gamma(\frac{n}{2})}\rho^{n-1}d\rho = \frac{2\pi^{\frac{n}{2}}}{n\Gamma(\frac{n}{2})}r^n = \frac{\pi^{\frac{n}{2}}r^n}{\Gamma(\frac{n}{2}+1)} \tag{1.255}$$

Explicitly

$$\Pr\{N(S_r^3) = k\} = e^{-\lambda\left(\frac{4}{3}\pi r^3\right)}\frac{\left[\lambda\left(\frac{4}{3}\pi r^3\right)\right]^k}{k!} \tag{1.256}$$

Clearly

$$\Pr\{N(S_r^2) = k\} = e^{-\lambda(\pi r^2)}\frac{[\lambda(\pi r^2)]^k}{k!} \tag{1.257}$$

If $D = \{$*distance of an arbitrary point* x *of the plane to the nearest neighbour*$\}$ we have that

$$\Pr\{D > r\} = \Pr\{N(S_r^2) = 0\} = e^{-\lambda\pi r^2} \tag{1.258}$$

and

$$\Pr\{D \in dr\} = f(r)dr = 2\lambda\pi r e^{-\lambda\pi r^2}dr, \qquad r > 0, \tag{1.259}$$

which is a Rayleigh distribution.

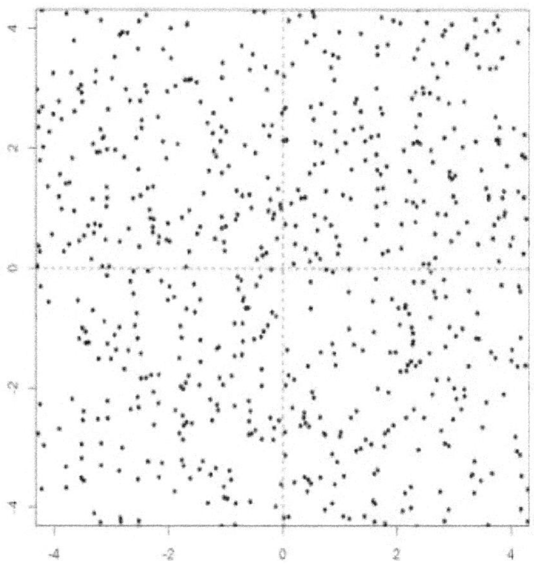

FIGURE 1.12
A general sample path of a Poisson random field in two dimensions.

Example. An example of a Poisson field is the distribution of flying bombs that were dropped over London during World War II (1940-1941). Clarke (1946) tested the spatial distribution, concluding that the Poisson law was confirmed and no spatial clustering existed.

1.7 The Integral of an Homogeneous Poisson Process

In this section we study the integral of a homogeneous Poisson process (see Orsingher and Polito, 2013):

$$\mathcal{N}(t) = \int_0^t N(s)ds = \sum_{j=2}^{n}(T_j - T_{j-1})(j-1) + n(t - T_n) \quad (1.260)$$

where, as in (1.19),

$$\Pr\{T_1 \in dt_1, \cdots, T_n \in dt_n \mid N(t) = n\} = \frac{n! dt_1 \cdots dt_n}{t^n}, \quad (1.261)$$

with $0 < t_1 < t_2 < \cdots < t_n < t$.

It is possible to show (see Theorem (1.17)) that the conditional *characteristic function* of $\mathcal{N}(t)$ reads

$$\mathbb{E}\left[e^{i\beta\mathcal{N}(t)}\middle|\mathcal{N}(t) = n\right] = \frac{(e^{i\beta t} - 1)^n}{t^n (i\beta)^n} \quad (1.262)$$

while the unconditional *characteristic function* is

$$\mathbb{E}\left(e^{i\beta\mathcal{N}(t)}\right) = e^{\lambda \int_0^t (e^{i\beta s} - 1) ds} \quad (1.263)$$

because

$$\begin{aligned}\mathbb{E}\left[e^{i\beta\mathcal{N}(t)}\right] &= \\ &= \mathbb{E}\left[e^{i\beta \int_0^t N(s)ds}\right] = \\ &= \sum_{n=0}^{\infty} \mathbb{E}\left[e^{i\beta \int_0^t N(s)ds} \mid N(t) = n\right] \cdot Pr\{N(t) = n\} = \\ &= \sum_{n=0}^{\infty} \frac{(e^{i\beta t} - 1)^n}{t^n (i\beta)^n} e^{-\lambda t} \frac{(\lambda t)^n}{n!} = e^{-\lambda t} \sum_{n=0}^{\infty} \frac{\left[\lambda(e^{i\beta t} - 1)\right]^n}{(i\beta)^n n!} = \\ &= e^{-\lambda t} e^{\frac{\lambda(e^{i\beta t} - 1)}{i\beta}} = e^{-\lambda t} e^{\lambda \int_0^t e^{i\beta s} ds} = \\ &= e^{\lambda \int_0^t (e^{i\beta s} - 1) ds}.\end{aligned} \quad (1.264)$$

Theorem 1.18. *For a homogeneous Poisson process $N(t)$, $t > 0$, we have that*

$$\mathcal{N}(t) = \int_0^t N(s) ds \stackrel{d}{=} \sum_{j=1}^{N(t)} X_j, \quad (1.265)$$

here the X_j's are i.i.d. random variables, Uniform in $[0, t]$. In 1.265, the sum on the right-hand side is intended to be equal to zero when $N(t) = 0$.

Proof. If $N(t) = n$, and $\tau_1, \tau_2, ..., \tau_n$ are the random instants at which the Poisson events appear, we have that the integral $A_n(t)$, $t \geq 0$, of the Poisson process reads

$$A_n(t) = \sum_{j=2}^{n}(\tau_j - \tau_{j-1})(j-1) + n(t - \tau_n). \tag{1.266}$$

Since

$$\Pr\{\tau_1 \in ds_1, ..., \tau_n \in ds_n\} = \frac{n!}{t^n}, \qquad 0 < s_1 < s_2 < ... < s_n < t, \tag{1.267}$$

we have

$$\begin{aligned} \mathbb{E}[e^{i\beta A_n(t)}] &= \\ &= \frac{n!}{t^n} \int_0^t ds_1 ... \int_{s_{n-1}}^t ds_n e^{i\beta\left[\sum_{j=2}^n (s_j - s_{j-1})(j-1) + n(t-s_n)\right]} = \\ &= \frac{n!}{t^n} F_n(\beta, t) \end{aligned} \tag{1.268}$$

It is evident that the functions $F_n(\beta, t)$ satisfy the equations

$$\begin{cases} \frac{d}{dt} F_n(\beta, t) = i\beta n F_n(\beta, t) + F_{n-1}(\beta, t), & n \geq 1, \\ F_n(\beta, 0) = 0, \end{cases} \tag{1.269}$$

where $F_0(\beta, t) = 0$. Now we show by induction that $F_n(\beta, t) = \frac{1}{n!}\left(\frac{(e^{i\beta t} - 1)}{i\beta}\right)^n$, $n \geq 0$. Hence, assuming that the expression holds for n, from (1.269) we have that

$$\frac{d}{dt} F_n(\beta, t) - i\beta n F_n(\beta, t) = \frac{1}{(n-1)!}\left(\frac{e^{i\beta t} - 1}{i\beta}\right)^{n-1} \tag{1.270}$$

and the solutions turn out to be

$$
\begin{aligned}
F_n(\beta,t) = e^{i\beta nt}\left[\int_0^t \frac{1}{(n-1)!}\left(\frac{e^{i\beta s}-1}{i\beta}\right)^{n-1} e^{-i\beta ns}ds\right] = \\
= \frac{e^{i\beta nt}}{(n-1)!(i\beta)^{n-1}}\sum_{m=0}^{n-1}\binom{n-1}{m}(-1)^{n-1-m}\int_0^t e^{i\beta(m-n)s}ds = \\
= \frac{e^{i\beta nt}}{(n-1)!(i\beta)^{n}}\sum_{m=0}^{n-1}\binom{n-1}{m}(-1)^{n-1-m}\left[\frac{e^{i\beta(m-n)t}-1}{(m-n)}\right] = \\
= \frac{e^{i\beta nt}}{(n-1)!(i\beta)^{n}}\sum_{m=0}^{n-1}\frac{(n-1)!}{m!(n-1-m)!}\frac{(-1)^{n-m}}{n-m}[e^{-i\beta(n-m)t}-1] = \\
= \frac{e^{i\beta nt}}{n!(i\beta)^{n}}\left[\sum_{m=0}^{n-1}\binom{n}{m}(-1)^{n-m}(e^{-i\beta(n-m)t}-1)\right] = \\
= \frac{e^{i\beta nt}}{n!(i\beta)^{n}}\left[\sum_{m=1}^{n}\binom{n}{m}(-1)^{m}(e^{-i\beta tm}-1)\right] = \\
= \frac{e^{i\beta nt}}{n!(i\beta)^{n}}\left[\sum_{m=0}^{n}\binom{n}{m}(-1)^{m}(e^{-i\beta tm}-1)\right] = \\
= \frac{e^{i\beta n}}{n!(i\beta)^{n}}\left[\sum_{m=0}^{n}\binom{n}{m}(-1)^{m}e^{-i\beta m} - \sum_{m=0}^{n}\binom{n}{m}(-1)^{m}\right] = \\
= \frac{e^{i\beta nt}}{n!(i\beta)^{n}}\left(1-e^{-i\beta t}\right)^n = \frac{(e^{i\beta t}-1)^n}{n!(i\beta)^n}
\end{aligned}
\tag{1.271}
$$

The *characteristic function* of $A_n(t)$ can thus be written as

$$
\mathbb{E}[e^{i\beta A_n(t)}] = \frac{(e^{i\beta t}-1)^n}{(i\beta)^n t^n}, \tag{1.272}
$$

so that

$$
\begin{aligned}
\mathbb{E}[e^{i\beta \mathcal{N}(t)}] = e^{-\lambda t}\sum_{n=0}^{\infty}(e^{i\beta t}-1)^n\frac{(\lambda t)^n}{n!(i\beta)^n)t^n} = e^{-\lambda t+\lambda\frac{e^{i\beta t}-1}{i\beta}} = \\
= e^{\lambda\int_0^t(e^{i\beta s}-1)ds},
\end{aligned}
\tag{1.273}
$$

which is the *characteristic function* of the compound process $\mathcal{N}(t)$, $t \geq 0$ (see eq. (1.204) as for the p.g.f.).

□

For small values of t the following approximation holds:

$$\mathbb{E}\left[e^{i\beta\mathcal{N}(t)}\right] = e^{i\lambda\beta\frac{t^2}{2}-\lambda\beta^2\frac{t^3}{6}+o(t^3)} \tag{1.274}$$

This implies that, for small value of t, the area below a homogeneous Poisson process is distributed as a Normal r.v. with mean $\mathbb{E}[\mathcal{N}(t)] = \frac{\lambda t^2}{2}$ and variance $\mathbb{V}\mathrm{ar}[\mathcal{N}(t)] = \frac{\lambda t^3}{3}$, since for a Normal distribution $N(\mu, \sigma^2)$ the *characteristic function* is $\mathbb{E}\left[e^{-i\beta N(\mu,\sigma^2)}\right] = e^{i\beta\mu-\beta^2\frac{\sigma^2}{2}}$.

Moreover

(i) the relationship

$$\mathbb{E}\left[\int_0^t N^k(s)ds \middle| N(t) = n\right] = \frac{t}{n+1}\sum_{j=1}^{n} j^k = \\ = \begin{cases} \frac{nt}{2} & k=1 \\ \frac{n(2n+1)t}{6} & k=2 \\ \frac{n^2(n+1)t}{4} & k=3 \end{cases} \tag{1.275}$$

holds, where we applied the following formulas

$$\begin{aligned} \sum_{j=1}^{n} j &= \frac{n(n+1)}{2} \\ \sum_{j=1}^{n} j^2 &= \frac{n(n+1)(2n+1)}{6} \\ \sum_{j=1}^{n} j^3 &= \left(\frac{n(n+1)}{2}\right)^2 = \left(\sum_{j=1}^{n} j\right)^2; \end{aligned} \tag{1.276}$$

(ii) by conditioning out (1.275) we easily have the unconditional values

$$\mathbb{E}\left[\int_0^t N(s)^k ds\right] = \begin{cases} \frac{\lambda t^2}{2} & k=1 \\ \frac{\lambda^2 t^3}{3} + \frac{\lambda t^2}{2} & k=2 \\ \frac{\lambda^3 t^4}{4} + \lambda^2 t^3 + \frac{\lambda t^2}{2} & k=3 \end{cases} \tag{1.277}$$

For the second order moment of (1.260), we have the following results

$$\begin{aligned} \mathbb{E}\left[\left(\int_0^\infty N(s)ds\right)^2 \middle| N(t)=n\right] &= \tfrac{n(3n+1)t^2}{12} \\ \mathbb{V}ar\left[\left(\int_0^\infty N(s)ds\right)^2 \middle| N(t)=n\right] &= \tfrac{nt^2}{12} \end{aligned} \tag{1.278}$$

as stated in Orsingher and Polito (2013). From (1.277) we extract the variance of the integral of the Poisson process

$$\begin{aligned} &\mathbb{V}ar\left[\int_0^t N(s)ds\right] = \\ &= \mathbb{E}\left[\mathbb{V}ar\left[\int_0^t N(s)ds|N(t)\right]\right] + \mathbb{V}ar\left[\mathbb{E}\left[\int_0^t N(s)ds|N(t)\right]\right] = \\ &= \frac{\lambda t^3}{3}. \end{aligned} \tag{1.279}$$

1.8 Iterated Poisson: Composition of Two Independent Poisson Processes

We now study the composition of two independent Poisson processes N_α and N_β of rates λ_α and λ_β (for details see Orsingher and Polito, 2012a).

The process (iterated Poisson)

$$\mathcal{M}(t) \equiv N_\alpha\left(N_\beta(t)\right) \tag{1.280}$$

is a Poisson process sampled at the random time $N_\beta(t)$.

Theorem 1.19. *The iterated process $\mathcal{M}(t)$ has distribution:*

$$Pr\left\{\mathcal{M}(t)=k\right\} = \frac{\lambda_\alpha^k}{k!} e^{-\lambda_\beta t\left(1-e^{-\lambda_\alpha}\right)} \underbrace{B_k\left(e^{-\lambda_\alpha}\lambda_\beta t\right)}_{\textit{Bell polynomials}} \tag{1.281}$$

where the Bell polynomials[6] *are defined as* $B_k(x) = e^{-x} \sum_{h=0}^{+\infty} \frac{h^k x^h}{h!}$.

Proof. In fact

$$\begin{aligned}
\Pr\{\mathcal{M}(t) = k\} &= \mathbb{E}[\Pr\{N_\alpha(N_\beta(t)) = k \mid N_\beta(t)\}] = \\
&= \sum_{h=0}^{\infty} \Pr\{N_\alpha(h) = k\} \cdot \Pr\{N_\beta(t) = h\} = \\
= \sum_{h=0}^{\infty} e^{-\lambda_\alpha h} \frac{(\lambda_\alpha h)^k}{k!} e^{-\lambda_\beta t} \frac{(\lambda_\beta t)^h}{h!} &= e^{-\lambda_\beta t} \frac{\lambda_\alpha^k}{k!} \sum_{h=0}^{\infty} e^{-\lambda_\alpha h} h^k \frac{(\lambda_\beta t)^h}{h!} = \\
&= e^{-\lambda_\beta t} \frac{\lambda_\alpha^k}{k!} \sum_{h=0}^{\infty} h^k \frac{(e^{-\lambda_\alpha} \lambda_\beta t)^h}{h!} = \\
&= e^{-\lambda_\beta t} \frac{\lambda_\alpha^k}{k!} e^{e^{-\lambda_\alpha} \lambda_\beta t} \cdot \underbrace{e^{-e^{-\lambda_\alpha} \lambda_\beta t} \sum_{h=0}^{\infty} h^k \frac{(e^{-\lambda_\alpha} \lambda_\beta t)^h}{h!}}_{\text{Bell polynomials } \mathcal{B}_k(e^{-\lambda_\alpha} \lambda_\beta t)} = \\
&= \frac{\lambda_\alpha^k}{k!} e^{-\lambda_\beta t (1 - e^{-\lambda_\alpha})} B_k(e^{-\lambda_\alpha} \lambda_\beta t)
\end{aligned} \tag{1.282}$$

□

The following relationship between the iterated Poisson $\mathcal{M}(t)$ and a special compound Poisson process (compound Poisson-Poisson) holds:

$$N_\alpha(N_\beta(t)) \overset{d}{\sim} \sum_{j=1}^{N_\beta(t)} N_j \tag{1.283}$$

where the N_j are i.i.d. Poisson distributed r.v.'s with parameter λ_α.

Thus $\mathcal{M}(t)$ can be regarded as a time-changed Poisson process and is a process with jumps of integer-valued random size.

[6]They are also known as Touchard polynomials or exponential Bell polynomials. They represent the moments of the Poisson distribution, i.e. $B_k(\lambda) = \sum_{n=0}^{\infty} n^k \cdot e^{-\lambda} \frac{\lambda^n}{n!}$, where λ is the intensity of the Poisson r.v. In R one can refer to the function *eBellPol* of the package *kStatistics*.

Remark 1.17.
Note further that this iterated Poisson process can count the total number of claims arising from a Poisson number of cars N_β having *each* a Poisson number of accidents N_j. It can also be noted that we can define a compound Poisson-Binomial model, $\sum_{j=1}^{N_\beta(t)} B_j(p)$ counting the total number of fatal crashes arising from a Poisson number of accidents N_β, where a Bernoulli variable $B_j(p)$ takes value 1 if the accident is fatal and zero if not.

Theorem 1.20. *The mean and variance of $\mathcal{M}(t)$ are*

$$\begin{aligned} \mathbb{E}[\mathcal{M}(t)] &= \lambda_\alpha \lambda_\beta t \\ \mathbb{V}ar[\mathcal{M}(t)] &= \lambda_\alpha(\lambda_\alpha + 1)\lambda_\beta t > \mathbb{E}[\mathcal{M}(t)] \end{aligned} \tag{1.284}$$

Proof. From the definition

$$\begin{aligned} \mathbb{E}[\mathcal{M}(t)] = \mathbb{E}[N_\alpha(N_\beta(t))] = \\ = \mathbb{E}_{N_\beta}[\mathbb{E}[N_\alpha(N_\beta(t)) \mid N_\beta(t)]] = \mathbb{E}[\lambda_\alpha N_\beta(t)] = \lambda_\alpha \lambda_\beta t; \\ \mathbb{V}ar[\mathcal{M}(t)] = \mathbb{E}[\mathcal{M}^2(t)] - (\mathbb{E}[\mathcal{M}(t)])^2 \end{aligned} \tag{1.285}$$

and

$$\begin{aligned} \mathbb{E}[\mathcal{M}^2(t)] = \mathbb{E}\left[N_\alpha^2(N_\beta(t))\right] = \mathbb{E}_{N_\beta}[\mathbb{E}[N_\alpha^2(N_\beta(t)) \mid N_\beta(t)]] = \\ = \sum_{h=0}^{\infty} \mathbb{E}\left[N_\alpha^2(h)\right] \Pr\{N_\beta(t) = h\} = \sum_{h=0}^{\infty} \left[\lambda_\alpha^2 h^2 + \lambda_\alpha h\right] e^{-\lambda_\beta t} \frac{(\lambda_\beta t)^h}{h!} = \\ = \lambda_\alpha^2 \sum_{h=0}^{\infty} h^2 e^{-\lambda_\beta t} \frac{(\lambda_\beta t)^h}{h!} + \lambda_\alpha \sum_{h=0}^{\infty} h e^{-\lambda_\beta t} \frac{(\lambda_\beta t)^h}{h!} = \\ = \lambda_\alpha^2 \left[\lambda_\beta^2 t^2 + \lambda_\beta t\right] + \lambda_\alpha \lambda_\beta t \end{aligned} \tag{1.286}$$

and

$$\begin{aligned} \mathbb{V}ar[\mathcal{M}(t)] = \lambda_\alpha^2 \lambda_\beta^2 t^2 + \lambda_\alpha^2 \lambda_\beta t + \lambda_\alpha \lambda_\beta t - \lambda_\alpha^2 \lambda_\beta^2 t^2 = \\ = \lambda_\alpha \lambda_\beta t(\lambda_\alpha + 1) \end{aligned} \tag{1.287}$$

□

Note that the computation of the variance is straightforward by applying formula (1.186) to the compound Poisson representation of the process, i.e.

$$\mathbb{V}ar[\mathcal{M}(t)] = \mathbb{V}ar[N_\beta(t)](\mathbb{E}[N_\alpha(t)])^2 + \mathbb{V}ar[N_\alpha(t)]\mathbb{E}[N_\beta(t)] = \\ = \lambda_\alpha\lambda_\beta(1+\lambda_\alpha)t \tag{1.288}$$

It is interesting to note that the iterated Poisson process is having the variance greater than the mean.

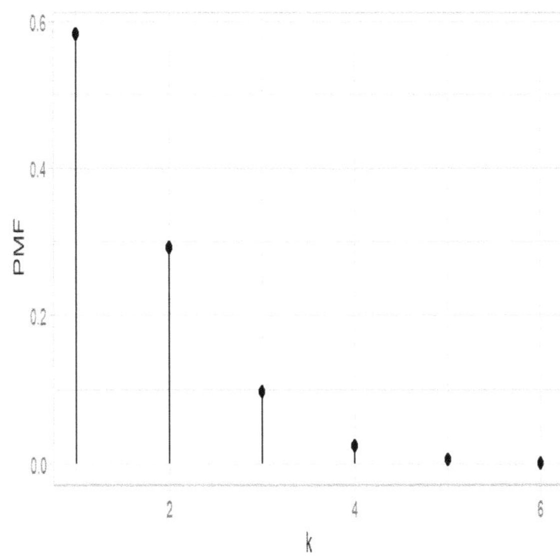

FIGURE 1.13
Iterated Poisson process: Probability mass function of the hitting time at time $t = 1$ against various levels k of upcrossing and $\lambda_\alpha = 1$.

For the iterated Poisson process, the distribution of the first-passage time through a level k, which is defined as

$$T_k = \inf\{s > 0 : \mathcal{M}(t) = N_\alpha(N_\beta(t)) = k\}, \quad k \geq 1 \tag{1.289}$$

is of great interest. In Orsingher and Polito (2012a), the authors derive the distribution of the hitting time T_k.

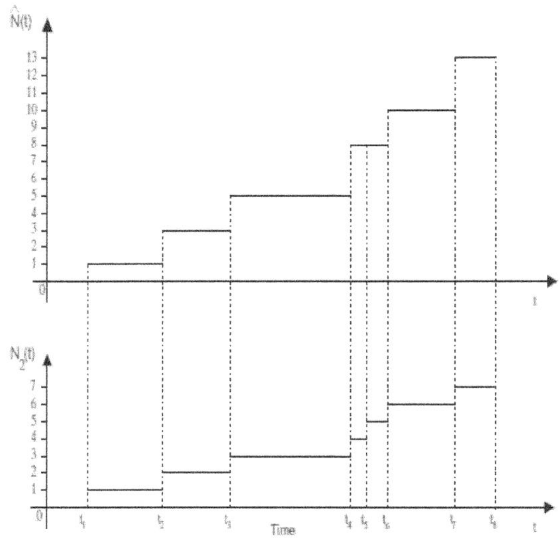

FIGURE 1.14
A sample path of the iterated Poisson process showing the composition of the inner and outer processes.

Theorem 1.21. *The distribution of the hitting time T_k of the iterated Poisson process, $k \geq 0$, reads as*

$$Pr\{T_k \in ds\} =$$

$$= e^{-\lambda_\alpha} \frac{\lambda_\alpha^k}{k!} \lambda_\beta e^{-\lambda_\beta s} ds \sum_{j=0}^{\infty} e^{-\lambda_\alpha j} [(j+1)^k - j^k] \frac{(\lambda_\beta s)^j}{j!}, \qquad s > 0. \tag{1.290}$$

Proof. First, the hitting time distribution is decomposed as

$$\Pr\{T_k \in ds\} =$$

$$= \sum_{h=1}^{k} \Pr\{N_\alpha(N_\beta(s)) = k - h, N_\alpha(N_\beta(s + ds) = k)\} =$$

$$= \sum_{h=1}^{k} \Pr\{N_\alpha(N_\beta(s)) = k - h, N_\alpha(N_\beta(s) + dN_\beta(s)) = k\} \tag{1.291}$$

Clearly $dN_\beta(s)$ takes either the value 0 with probability $1-\lambda_\beta ds$ or the value 1. Therefore

$$\Pr\{T_k \in ds\} =$$

$$= \lambda_\beta ds \sum_{h=1}^{k} \Pr\{N_\alpha(N_\beta(s)) = k-h, N_\alpha(N_\beta(s)+1) = k\} \tag{1.292}$$

In the interval $(N_\beta(s), N_\beta(s)+1)$, the external process can take all possible values $0 \leq k-h \leq k-1$. The value k must be excluded because if at time s, $N_\alpha(N_\beta(s)) = k$, the first attainment of value k cannot be recorded during the interval $(s, s+ds]$. Furthermore,

$$\begin{aligned}\Pr\{N_\alpha(N_\beta(s)) = k-h, N_\alpha(N_\beta(s)+1) = k \mid N_\beta(s) = j\} &= \\ = \Pr\{N_\alpha(j) = k-h, N_\alpha(j+1) = k\} &= \\ = \Pr\{N_\alpha(j) = k-h\} \times \Pr\{N_\alpha(j+1) - N_\alpha(j) = h\} &= \\ = e^{\lambda_\alpha j}\frac{(\lambda_\alpha j)^{k-h}}{(k-h)!} e^{-\lambda_\alpha}\frac{\lambda_\alpha^h}{h!}&\end{aligned} \tag{1.293}$$

By substituting equation (1.293) into (1.291), the result obtains, i.e.

$$\begin{aligned}\Pr\{T_k \in ds\} &= \\ &= ds\lambda_\beta e^{-\lambda_\beta s} \times \\ &\times \sum_{j=0}^{\infty}\sum_{h=1}^{k}\left[e^{-\lambda_\alpha j}\frac{(\lambda_\alpha j)^{k-h}}{(k-h)!}e^{-\lambda_\alpha}\frac{\lambda_\alpha^h}{h!}\right]\cdot\frac{(\lambda_\beta s)^j}{j!} = \\ &= ds\lambda_\beta e^{-\lambda_\beta s}e^{-\lambda_\alpha}\frac{\lambda_\alpha^k}{k!}\sum_{j=0}^{\infty} e^{-\lambda_\alpha j}\left[(j+1)^k - j^k\right]\frac{(\lambda_\beta s)^j}{j!},\end{aligned} \tag{1.294}$$

since

$$\begin{aligned}\frac{1}{k!}\sum_{h=1}^{k}\binom{k}{h}(\lambda_\alpha j)^{k-h}\lambda_\alpha^h = \frac{1}{k!}(\lambda_\alpha j + \lambda_\alpha)^k - (\lambda_\alpha j)^k &= \\ = \frac{\lambda_\alpha^k}{k!}\left[(j+1)^k - j^k\right].&\end{aligned} \tag{1.295}$$

□

For the hitting probabilities $\Pr\left(T_k^{(\alpha,\beta)} < \infty\right)$ we have that

$$\begin{aligned}
\Pr\left(T_k^{(\alpha,\beta)} < \infty\right) &= \int_0^\infty \Pr\{T_k^{(\alpha,\beta)} \in ds\} = \\
&= e^{-\lambda_\alpha} \frac{\lambda_\alpha^k}{k!} \sum_{j=0}^\infty e^{-\lambda_\alpha j} \left[(j+1)^k - j^k\right] = \\
&= \frac{\lambda_\alpha^k}{(k-1)!} \sum_{j=0}^\infty \int_j^{j+1} e^{-\lambda_\alpha (j+1)} x^{k-1} dx < \\
&< \frac{\lambda_\alpha^k}{(k-1)!} \sum_{j=0}^\infty \int_j^{j+1} e^{-\lambda_\alpha x} x^{k-1} dx = \\
&= \frac{\lambda_\alpha^k}{(k-1)!} \int_0^\infty e^{-\lambda_\alpha x} x^{k-1} dx = 1, \quad \forall k
\end{aligned} \tag{1.296}$$

which is independent of the intensity of the inner process λ_β and, in particular,

$$\Pr\{T_1 < \infty\} = \lambda_\alpha e^{-\lambda_\alpha} \sum_{j=0}^\infty e^{-\lambda_\alpha j} = \frac{\lambda_\alpha e^{-\lambda_\alpha}}{1-e^{-\lambda_\alpha}} < 1. \tag{1.297}$$

It means that there is a positive probability of never reaching level k for the iterated Poisson process.

1.9 Shot Noise and Hawkes Processes

We introduce shot noise processes and the Hawkes process as a particular instance.

1.9.1 The Shot Noise Process

The notion of shot noise (process) or *Poisson shot noise* was introduced around 1919 by Walter Schottky and is defined as

$$Y(t) = \sum_{i=1}^{N(t)} h(t - \tau_i) \tag{1.298}$$

where $\tau_1 < ... < \tau_i < ... < \tau_n < t$ are time instants of n occurrences of a homogeneous Poisson process in a time interval $(0, t)$.

Therefore, $Y(t)$ represents the effect at time t of all events happened at times $\tau_1, ..., \tau_n$, and *filtered* through the *response function* $h(\cdot)$.

The shot noise can be extended by introducing the *random* impulses X_i of the events at times τ_i as

$$Y(t) = \sum_{i=1}^{N(t)} X_i \cdot h(t - \tau_i) \tag{1.299}$$

the X_i's being i.i.d. random variables.

The *shot noise* with an exponential response function can be represented as a random sum, whose summands are i.i.d. Exponential r.v.'s of parameter δ and with a Binomial number of summands $M(t)$, with parameters $(n, 1 - e^{-\beta t})$, i.e.

$$Y(t) = X_1 + ... + X_{M(t)} \tag{1.300}$$

When the response function is exponential, no matter the distribution of impulses, the process $Y(t)$ is Markovian, since for $s < t$:

$$Y(t) = e^{-\beta(t-s)} Y(s) + \sum_{i=N(s)+1}^{N(t)} X_i e^{-\beta(t-\tau_i)} \tag{1.301}$$

See also Orsingher, Battaglia (1982) for some shot noise models driven by various point processes.

1.9.2 Hawkes Processes

Definition. A Hawkes process $N^H(t)$ is a Poisson process driven by a *conditional intensity* $\tilde{\lambda}(t)$ subject to *shot noise*. As such, it is a case of a shot noise process, where the shot noise is deterministic.

More precisely, in Hawkes processes the shot noise acts on the conditional intensity $\tilde{\lambda}_t$ as follows:

$$\tilde{\lambda}_t = \mu + (\lambda_0 - \mu)e^{-\beta t} + \alpha \sum_{j=1}^{N^H(t)} e^{-\beta(t-\tau_j)} \tag{1.302}$$

where the shot noise α is a deterministic shock or "impulse", that increases (for $\alpha > 0$, one speaks of a *self-reinforcing* mechanism) the intensity at time points τ_i, with a temporary effects that decays exponentially at a decay rate β. Moreover, μ represents a *baseline intensity* that can be seen as a *floor* level, while $\lambda_0 > \mu$ is the starting point of the intensity ($\lambda_0 = \mu$ highlights the absence of a previous history).

In a Hawkes process $N^H(t)$, the *intensity* function $\tilde{\lambda}(t)$ depends on the past history of the process: it is subject to increases in case of Poisson events (occurring at random instants $\tau_i > 0$, $i \geq 1$) while it is fading away between them.

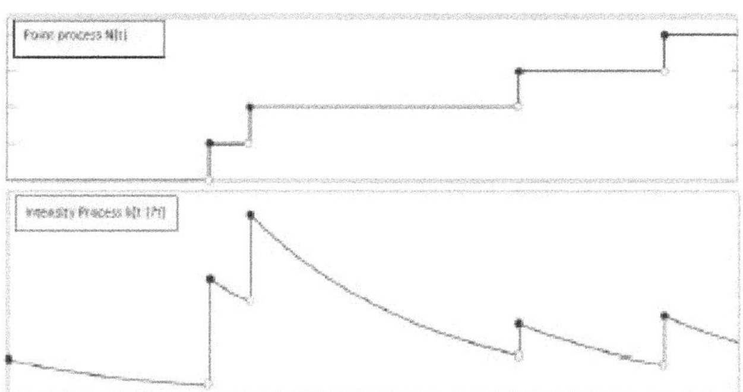

FIGURE 1.15
Description of a Hawkes process: the point process and its driving intensity (from Laub, Lee, and Taimre, 2022).

In the context of Hawkes processes, the intensity is written in terms of an *excitation function* $h(t) = \alpha e^{-\beta t}$ and it also reads as

$$\tilde{\lambda}(t) = \mu + \int_0^t h(t-s)dN(s) = \mu + \sum_{\tau_i < t} h(t - \tau_i) \tag{1.303}$$

Example. As a case study, a Hawkes process has been estimated on (*night*[7]) accidents that occurred on two urban roads in the city of Rome in the year 2019 (the last year before the pandemic), as provided by local police report data.

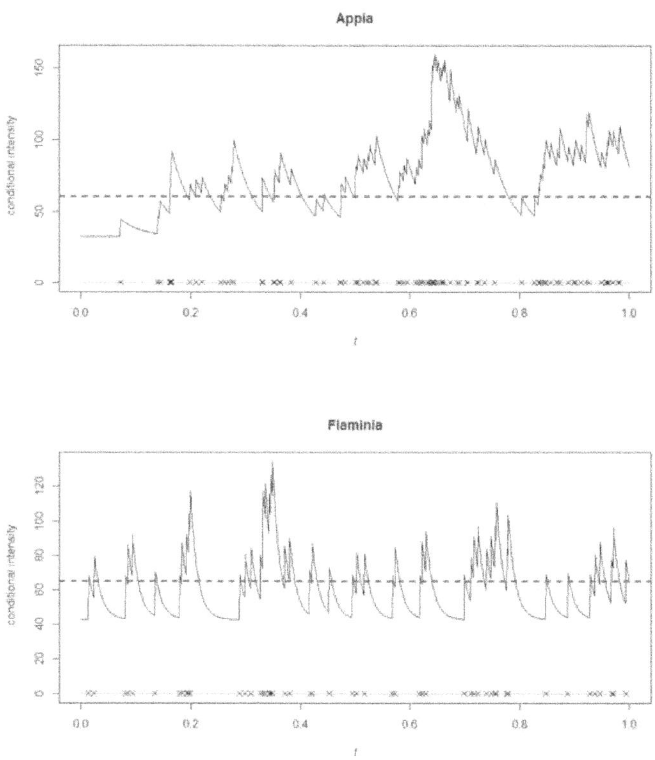

FIGURE 1.16
Conditional intensity of a Hawkes process estimated on (*night*) vehicle accidents data, occurred in the municipality of Rome on two *urban* roads (*Via Appia* and *Via Flaminia*), in 2019.

[7]Night was defined as accidents occurred between 1 a.m. and 6 a.m.

The parameters' values are estimated[8] as (over $T = 1$ year):

- Appia: $\hat{\mu} = 32.3$, $\hat{\beta} = 26.7$, $\hat{\alpha} = 12.4$
- Flaminia: $\hat{\mu} = 42.9$, $\hat{\beta} = 77.3$, $\hat{\alpha} = 26.4$

[8]The estimation has been carried out by means of the R package '*howkes-bow*', assuming $\lambda_0 = \mu$.

Part II

Birth-and-death processes

2

Birth processes

CONTENTS

This chapter covers the *pure-birth process* in its most common versions. The pure-birth process is introduced starting from the non linear case, by stating the difference-differential equation for the state distribution $p_k(t)$ and the related Cauchy problem whose solution is obtained both by solving the ODE and by Laplace transform. Two properties of this process are presented: i) the relation between the series of the reciprocals of the process rates and the probability of explosion and ii) the Exponential distribution of the inter-event times.

Then for the classical linear process (a.k.a. Yule-Furry model), the size distribution is obtained as a Geometric distribution in the case of 1 progenitor and as a Binomial distribution in the case of n_0 progenitors. The process p.g.f. is derived as a solution of a PDE. From the p.g.f. the expected value and variance obtain, confirming the exponential growth feature of the process. Moreover, the probability of "explosion" is also discussed.

Finally, the process inference is discussed by presenting classic results on the process likelihood.

DOI: 10.1201/9781003398011-2

2.1 The Nonlinear Case

A model for which the probability of new offspring depends on the current population size is now presented. We assume, for the sake of simplicity, that at time $t = 0$ there is only one progenitor.

Let $\mathcal{N}(t)$, $t > 0$, be the size of the population at time t; its probabilistic evolution is governed by the following rules ($k \geq 1$):

1. $\Pr\{\mathcal{N}(0) = 1\}=1$;
2. $\Pr\{\mathcal{N}(t, t + dt] = 1 \mid \mathcal{N}(t) = k\} = \lambda_k dt + o(dt)$;
3. $\Pr\{\mathcal{N}(t, t + dt] = 0 \mid \mathcal{N}(t) = k\} = 1 - \lambda_k dt + o(dt)$;
4. $\Pr\{\mathcal{N}(t, t + dt] > 1 \mid \mathcal{N}(t) = k\} = o(dt)$,

where $\lambda_k > 0$ are the time-independent birth-rates.

Remark 2.1.
The birth process obtained by assuming that the birth rates are time dependent, i.e. $\lambda_k = \lambda_k(t)$ is referred to as the *general* birth process (see Remark 4.2).

Remark 2.2.
Note that if $\lambda_k = \lambda \ \forall k$ the increments are independent and the process is a homogeneous Poisson process.

Note that the independence of increments, in this case, is no longer holding (see Fig. 2.1).

The state probabilities are:

$$p_k(t) = Pr\left\{\mathcal{N}(t) = k \mid \mathcal{N}(0) = 1\right\}, \tag{2.1}$$

satisfy the difference-differential equation

$$\frac{dp_k(t)}{dt} = -\lambda_k p_k(t) + \lambda_{k-1} p_{k-1}(t) \qquad t > 0 \tag{2.2}$$

with initial conditions

$$p_k(0) = \begin{cases} 1 & k = 1 \qquad \text{(one initial progenitor)} \\ 0 & k \geq 2 \end{cases} \tag{2.3}$$

The solution of (2.2)–(2.3) is obtained by means of a recursive procedure. If $k = 1$, the solution of

$$\begin{cases} \frac{dp_1(t)}{dt} = -\lambda_1 p_1(t), \\ p_1(0) = 1, \end{cases} \tag{2.4}$$

is $p_1(t) = e^{-\lambda_1 t}$. If $k = 2$, $p_2(t)$ it obtained by solving the non-homogeneous linear equation

$$\begin{cases} \frac{dp_2(t)}{dt} = -\lambda_2 p_2(t) + \lambda_1 e^{-\lambda_1 t}, \\ p_2(0) = 0, \end{cases} \tag{2.5}$$

and it reads

$$p_2(t) = e^{-\lambda_2 t}\left\{\int_0^t \lambda_1 e^{-\lambda_1 s} e^{\lambda_2 s}\right\} = \lambda_1 \left[\frac{e^{-\lambda_1 t}}{\lambda_2 - \lambda_1} + \frac{e^{-\lambda_2 t}}{\lambda_1 - \lambda_2}\right] \tag{2.6}$$

In order to understand how the distribution $p_k(t)$ emerges, we consider the case $k = 3$, that is

$$\begin{cases} \frac{dp_3(t)}{dt} = -\lambda_3 p_3(t) + \lambda_2 \lambda_1 \left[\frac{e^{-\lambda_1 t}}{\lambda_2 - \lambda_1} + \frac{e^{-\lambda_2 t}}{\lambda_1 - \lambda_2}\right], \\ p_3(0) = 0. \end{cases} \tag{2.7}$$

By proceeding as for (2.5) we have that

$$
\begin{aligned}
p_3(t) &= e^{-\lambda_3 t}\left[\int_0^t \lambda_2\lambda_1\left(\frac{e^{-\lambda_1 s}}{\lambda_2-\lambda_1}+\frac{e^{-\lambda_2 s}}{\lambda_1-\lambda_2}\right)e^{\lambda_3 s}ds\right] = \\
&= e^{-\lambda_3 t}\lambda_1\lambda_2\left[\frac{e^{(\lambda_3-\lambda_1)t}-1}{(\lambda_3-\lambda_1)(\lambda_2-\lambda_1)}+\frac{e^{(\lambda_3-\lambda_2)t}-1}{(\lambda_1-\lambda_2)(\lambda_3-\lambda_2)}\right] = \\
&= \lambda_1\lambda_2\left[\frac{e^{-\lambda_1 t}}{(\lambda_3-\lambda_1)(\lambda_2-\lambda_1)}+\frac{e^{-\lambda_2 t}}{(\lambda_1-\lambda_2)(\lambda_3-\lambda_2)}+\right. \\
&\left.-\frac{e^{-\lambda_3 t}}{\lambda_2-\lambda_1}\left(\frac{1}{\lambda_3-\lambda_1}-\frac{1}{\lambda_3-\lambda_2}\right)\right] = \lambda_1\lambda_2\left[\frac{e^{-\lambda_1 t}}{(\lambda_3-\lambda_1)(\lambda_2-\lambda_1)}+\right. \\
&\left.+\frac{e^{-\lambda_2 t}}{(\lambda_1-\lambda_2)(\lambda_3-\lambda_2)}+\frac{e^{-\lambda_3 t}}{(\lambda_1-\lambda_3)(\lambda_2-\lambda_3)}\right] \\
&= \prod_{j=1}^{2}\lambda_j\left[\sum_{m=1}^{3}\frac{e^{-\lambda_m t}}{\prod_{l=1,l\neq m}^{3}(\lambda_l-\lambda_m)}\right].
\end{aligned} \tag{2.8}
$$

Formula (2.8) suggests that the general expression for the probabilities $p_k(t)$ has the form

$$
p_k(t) = \begin{cases} \prod_{j=1}^{k-1}\lambda_j \sum\limits_{m=1}^{k}\frac{e^{-\lambda_m t}}{\prod_{l=1,l\neq m}^{k}(\lambda_l-\lambda_m)}, & k>1, \\ e^{-\lambda_1 t}, & k=1. \end{cases} \tag{2.9}
$$

Proceeding by induction, we must solve the following Cauchy problem

$$
\begin{cases} \frac{dp_k(t)}{dt} = -\lambda_k p_k(t) + \lambda_{k-1}\prod_{j=1}^{k-2}\lambda_j \sum\limits_{m=1}^{k-1}\frac{e^{-\lambda_m t}}{\prod_{l=1,l\neq m}^{k}(\lambda_l-\lambda_m)}, & k>1, \\ p_k(0)=0. \end{cases} \tag{2.10}
$$

The solution to (2.10) is obtained by applying once again the formula for the solution of first-order linear non-homogeneous equations[1] which yields:

$$\begin{aligned}
p_k(t) &= e^{-\lambda_k t}\left[\int_0^t \prod_{j=1}^{k-1}\lambda_j \sum_{m=1}^{k-1} \frac{e^{-\lambda_m s}}{\prod_{l=1,l\neq m}^{k-1}(\lambda_l-\lambda_m)} e^{\lambda_k s} ds\right] \\
&= e^{-\lambda_k t}\left[\prod_{j=1}^{k-1}\lambda_j \sum_{m=1}^{k-1} \frac{e^{(\lambda_k-\lambda_m)t}-1}{\prod_{l=1,l\neq m}^{k-1}(\lambda_l-\lambda_m)(\lambda_k-\lambda_m)}\right] \\
&= \prod_{j=1}^{k-1}\lambda_j \left[\sum_{m=1}^{k-1} \frac{e^{-\lambda_m t}}{\prod_{l=1,l\neq m}^{k}(\lambda_l-\lambda_m)} + \right. \qquad (2.11) \\
&\left. -e^{-\lambda_k t}\sum_{m=1}^{k-1} \frac{1}{\prod_{l=1,l\neq m}^{k}(\lambda_l-\lambda_m)}\right] \\
&= \prod_{j=1}^{k-1}\lambda_j \left[\sum_{m=1}^{k} \frac{e^{-\lambda_m t}}{\prod_{l=1,l\neq m}^{k}(\lambda_l-\lambda_m)}\right],
\end{aligned}$$

since[2]

$$-\sum_{m=1}^{k-1} \frac{1}{\prod_{l=1,l\neq m}^{k}(\lambda_l-\lambda_m)} = \frac{1}{\prod_{l=1,l\neq k}^{k}(\lambda_l-\lambda_k)} \qquad (2.12)$$

In order to retrieve $p_3(t)$ of formula (2.8) set $k = 3$ in (2.9), see Orsingher and Polito (2010).

---

[1]It is trivial to note that the terms correspond to the form $y'(t) + a(t)y(t) = b(t)$, where, for all $k > 1$, $a(t) \equiv \lambda_k$ and $b(t) \equiv \lambda_{k-1}\prod_{j=1}^{k-2}\lambda_j \sum_{m=1}^{k-1}\frac{e^{-\lambda_m t}}{\prod_{l=1,l\neq m}^{k}(\lambda_l-\lambda_m)}$.

[2]The result obtains after application of the Laplace expansion for determinants.

Remark 2.3.
From (2.2) we obtain the recursive relationship

$$\mathcal{L}[p_k](\mu) = \frac{\lambda_{k-1}}{\mu+\lambda_k}\mathcal{L}[p_{k-1}](\mu) \tag{2.13}$$

where $\mathcal{L}[p_k](\mu)$ is the Laplace transform of the distribution $p_k(t)$.

By developing (2.13) we get

$$\begin{aligned}\mathcal{L}[p_k](\mu) = \frac{\lambda_{k-1}}{\mu+\lambda_k}\frac{\lambda_{k-2}}{\mu+\lambda_{k-1}}\cdots\frac{\lambda_1}{\mu+\lambda_2}\frac{1}{\mu+\lambda_1} &= \\ = \prod_{j=1}^{k-1}\lambda_j \cdot \prod_{j=1}^{k}\frac{1}{\mu+\lambda_j}\end{aligned} \tag{2.14}$$

because $\int_0^\infty e^{-\mu t}p_1(t)dt = \int_0^\infty e^{-\mu t}e^{-\lambda_1 t}dt$.

By evaluating the Laplace transform of (2.9) we have

$$\begin{aligned}\int_0^\infty e^{-\mu t}p_k(t)dt &= \\ &= \prod_{j=1}^{k-1}\lambda_j \cdot \sum_{m=1}^{k}\frac{\int_0^\infty e^{-\mu t}e^{-\lambda_m t}dt}{\prod_{l=1,l\neq m}^{k}(\lambda_l-\lambda_m)} = \\ &= \prod_{j=1}^{k-1}\lambda_j \cdot \sum_{m=1}^{k}\frac{1}{\prod_{l=1,l\neq m}^{k}(\lambda_l-\lambda_m)}\cdot\frac{1}{\mu+\lambda_m}\end{aligned} \tag{2.15}$$

By comparing the result obtained in (2.14) from the recursive application of the Laplace transform with the Laplace transform of the distribution (2.13), we have that

$$\prod_{j=1}^{k}\frac{1}{\mu+\lambda_j} = \sum_{m=1}^{k}\frac{1}{\prod_{l=1,l\neq m}^{k}(\lambda_l-\lambda_m)}\frac{1}{\mu+\lambda_m} \tag{2.16}$$

Remark 2.4.
If $\sum_{m=1} \frac{1}{\lambda_m} = \infty$ then $\Pr\{\mathcal{N}(t) < \infty\} = 1$, so that the size population cannot explode in a finite time. This means that the birth rates must increase slowly so that the sequence $1/\lambda_m$, $m \geq 1$, forms a divergent series.

Theorem 2.1: The time between the k-th and $(k+1)$-th birth has distribution

$$\Pr\{T_k \in ds\} = \lambda_k e^{-\lambda_k s} ds, \qquad s > 0. \tag{2.17}$$

Proof. Let $Z_k = T_1 + \cdots + T_k$ be the waiting time for k-th birth. We have the following relation:

$$\Pr\{T_1 + \cdots + T_k \in dt\} = \\ = \int_0^t \Pr\{T_1 + \cdots + T_{k-1} \in ds\} \Pr\{T_k \in d(t-s)\}, \tag{2.18}$$

where

$$\Pr\{T_1 + \cdots + T_{k-1} \in ds\}/ds = \frac{d}{ds}\Pr\{\mathcal{N}(s) \geq k\}. \tag{2.19}$$

Consider that $\mathcal{N}$ includes also the progenitor[3].

By applying the Laplace transform to both members of (2.18), we have that

$$\begin{aligned} &\int_0^\infty e^{-\mu t} \Pr\{T_1 + \cdots + T_k \in dt\} \\ &= \int_0^\infty e^{-\mu t} dt \int_0^t \Pr\{T_k \in d(t-s)\} \Pr\{T_1 + \cdots + T_{k-1} \in ds\} \\ &= \int_0^\infty \Pr\{T_1 + \cdots + T_{k-1} \in ds\} \int_s^\infty e^{-\mu t} \Pr\{T_k \in d(t-s)\} \\ &= \int_0^\infty e^{-\mu s} \Pr\{T_1 + \cdots + T_{k-1} \in ds\} \int_0^\infty e^{-\mu y} \Pr\{T_k \in dy\} \\ &= \prod_{j=1}^k \int_0^\infty e^{-\mu s} \Pr\{T_j \in ds\} = \prod_{j=1}^k \frac{\lambda_j}{\mu + \lambda_j}. \end{aligned} \tag{2.20}$$

Each factor in (2.20) is the Laplace transform of (2.11).

[3] i.e. k includes one progenitor and $k-1$ offsprings.

Furthermore,

$$\Pr\{T_1 < s\} = \Pr\{N(s) \geq 2\} \tag{2.21}$$

and taking the derivative

$$\begin{aligned}\frac{\Pr\{T_1 \in ds\}}{ds} &= \frac{d}{ds}\Pr\{N(s) \geq 2\} = \frac{d}{ds}(1 - \Pr\{N(s) = 1\}) \\ &= \frac{d}{ds}(1 - e^{-\lambda_1 s}) = \lambda_1 e^{-\lambda_1 s}\end{aligned} \tag{2.22}$$

and

$$\begin{aligned}\Pr\{T_1 + T_2 < s\} = \Pr\{N(s) \geq 3\} = \\ = 1 - \Pr\{N(s) = 1\} - \Pr\{N(s) = 2\}\end{aligned} \tag{2.23}$$

and deriving:

$$\begin{aligned}\Pr\{T_1 + T_2 \in ds\}/ds &= \frac{d}{ds}\{1 - \Pr\{N(s) = 1\} - \Pr\{N(s) = 2\}\} = \\ &= \lambda_1 e^{-\lambda_1 s} + \lambda_1 \left[\lambda_1 \frac{e^{-\lambda_1 s}}{\lambda_2 - \lambda_1} + \lambda_2 \frac{e^{-\lambda_2 s}}{\lambda_1 - \lambda_2}\right] = \\ &= \frac{\lambda_1 \lambda_2}{\lambda_2 - \lambda_1}\left(e^{-\lambda_1 s} - e^{-\lambda_2 s}\right).\end{aligned} \tag{2.24}$$

Note that (2.20), for $k = 2$, becomes

$$\int_0^\infty e^{-\mu t} \Pr\{T_1 + T_2 \in dt\} = \frac{\lambda_1 \lambda_2}{(\lambda_1 + \mu)(\lambda_2 + \mu)}, \tag{2.25}$$

and corresponds to the Laplace transform of (2.24).

2.2 The Linear Case (Yule-Furry)

In this section we denote with $\hat{N}(t)$, $t > 0$, the size of the population at time t of a linear birth process (called also Yule-Furry

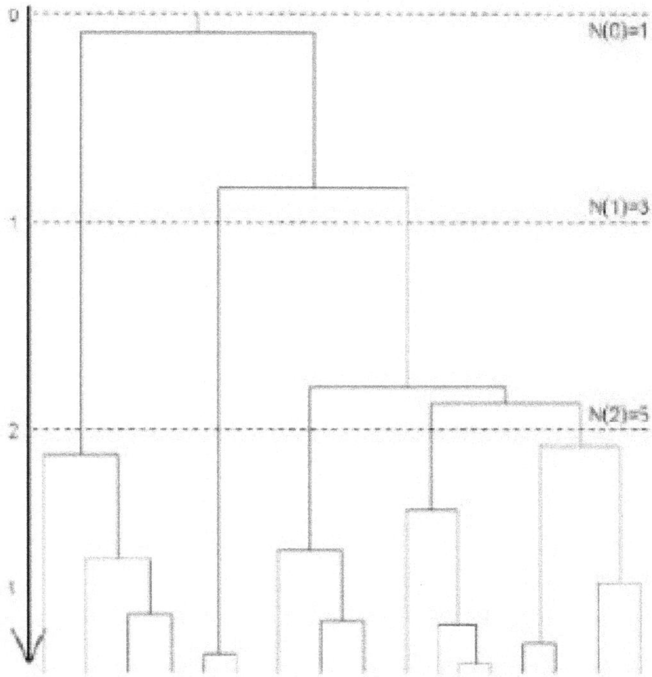

FIGURE 2.1
A general sample path of a birth processes.

process). In this case the birth rates become $\lambda_k = \lambda k$, with $\lambda > 0$. The distribution (2.9) becomes relatively simple. By observing that

$$\prod_{l=1, l \neq m}^{k} (l\lambda - m\lambda) = \lambda^{k-1} \times$$
$$\times [(1-m)(2-m)\dots(m-1-m)(m+1-m)\dots(k-m)] =$$
$$= \lambda^{k-1}(-1)^{m-1}(m-1)!(k-m)!\,, \tag{2.26}$$

formula (2.9) can be rewritten as

$$\begin{aligned}
p_k(t) &= \lambda^{k-1}(k-1)!\sum_{m=1}^{k}\frac{e^{-\lambda mt}}{\lambda^{k-1}(m-1)!(k-m)!}(-1)^{m-1} = \\
&= \sum_{m=1}^{k} e^{-\lambda mt}(-1)^{m-1}\frac{(k-1)!}{(m-1)!(k-m)!} = \\
&= \sum_{m=0}^{k-1} e^{-\lambda t(m+1)}(-1)^{m}\frac{(k-1)!}{m!(k-m-1)!} = \\
&= e^{-\lambda t}\sum_{m=0}^{k-1}\binom{k-1}{m}(-1)^m e^{-\lambda tm} = e^{-\lambda t}\left(1-e^{-\lambda t}\right)^{k-1}, k \geq 1,
\end{aligned} \tag{2.27}$$

which is a Geometric distribution with parameter $e^{-\lambda t}$ (see Fig. 2.2).

Remark 2.5.
Note that if the progenitors are $n_0 > 1$ the random number $\hat{N}(t)$ of components of the population becomes a negative Binomial distribution (the sum of n_0 independent Geometric r.v.'s):

$$p_k(t) = \binom{k-1}{k-n_0}e^{-\lambda t n_0}(1-e^{-\lambda t})^{k-n_0}, \qquad k \geq n_0,$$

(see Fig. 2.3 and Fig. 2.4).

Remark 2.6.
Yule (1924) (a British statistician) introduced the pure-birth process to study the emergence of new species. The process is sometimes referred to as the process of Yule-Furry as Furry (1937) (an American physicist) applied the process to the offsprings of radioactive particles.

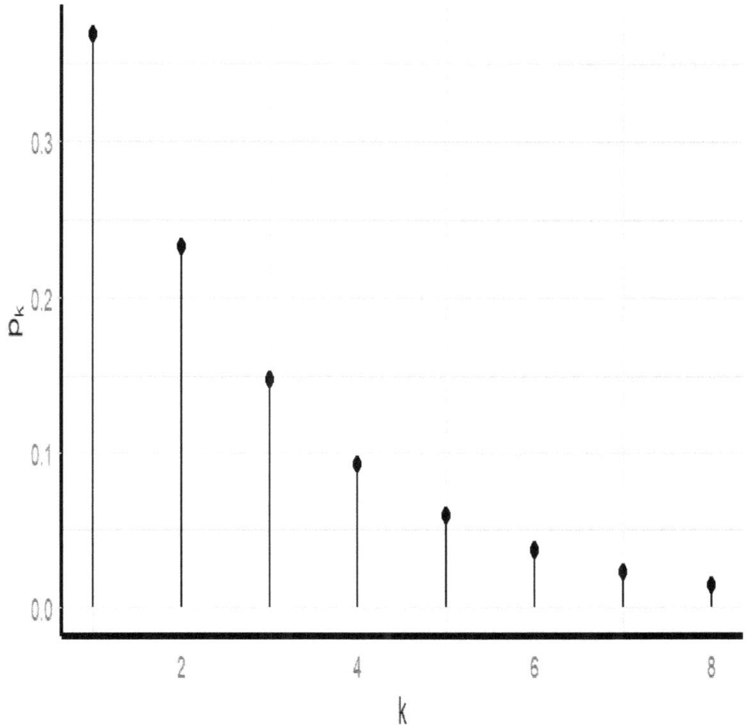

FIGURE 2.2
State probabilities $\{p_k(t),\ k \geq 1\}$ of a pure birth process (linear case), with one progenitor ($n_0 = 1$) and $\lambda = 1$; $t = 1$.

The probabilities (2.27) can be obtained directly by solving the following equations, recursively

$$\frac{dp_k(t)}{dt} = -\lambda k p_k(t) + \lambda(k-1)p_{k-1}(t) \tag{2.28}$$

with initial conditions

$$p_k(0) = \begin{cases} 1, & k = 1, \\ 0, & k \geq 2. \end{cases} \tag{2.29}$$

By using the probability generating function

$$G(u,t) = \sum_{k=1}^{\infty} u^k p_k(t), \quad |u| \leq 1 \tag{2.30}$$

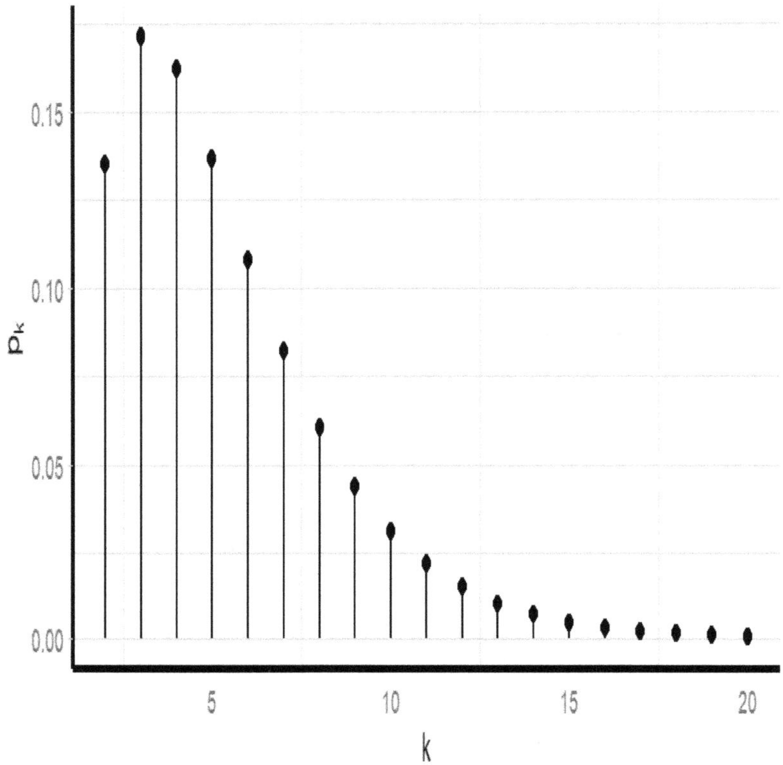

FIGURE 2.3
State probabilities $\{p_k(t), k \geq 1\}$ of a pure birth process (linear case), for $n_0 = 2$ progenitors (Negative Binomial) and $\lambda = 1$; $t = 1$.

we can pass from (2.28) to the following partial differential equation

$$\frac{\partial G(u,t)}{\partial t} = \lambda u(u-1)\frac{\partial}{\partial u}G(u,t) \tag{2.31}$$

with initial condition $G(u, 0) = u$.
The general solution of the first-order partial differential equation (2.31) is

$$G(u,t) = f\left(e^{-\lambda t}\frac{u}{1-u}\right) \tag{2.32}$$

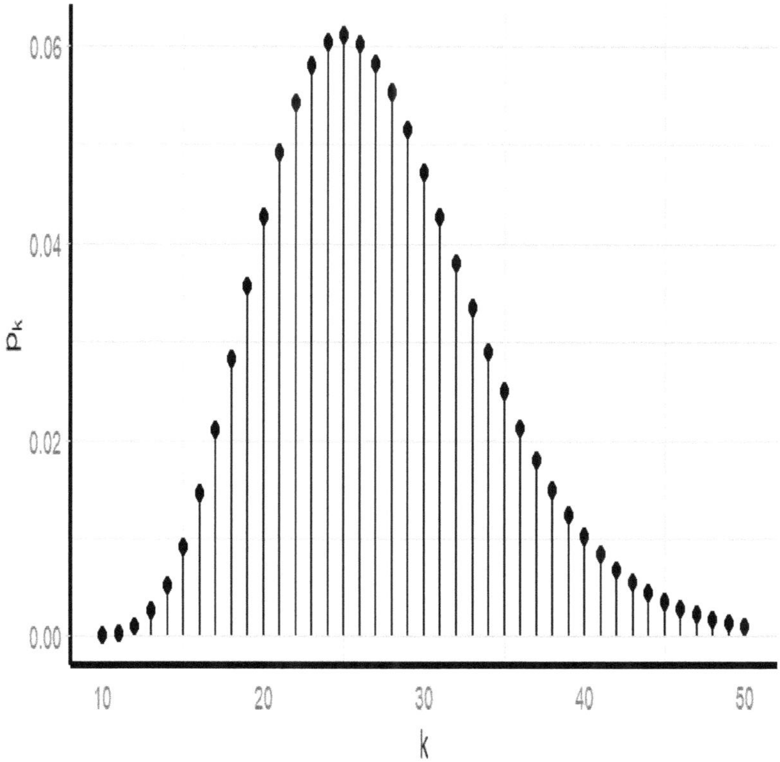

FIGURE 2.4
State probabilities $\{p_k(t), k \geq 1\}$ of a pure birth process (linear case), $n_0 = 10$ progenitors (negative binomial) and $\lambda = 1$; $t = 1$.

where $f \in \mathbb{C}^1$ is a differentiable function. The initial condition permits us to obtain the explicit form of f because

$$G(u, 0) = u = f\left(\frac{u}{1-u}\right) \tag{2.33}$$

and by setting $\frac{u}{1-u} = v$ we see that $f(v) = \frac{v}{1+v}$. From (2.32) we then have that

$$\begin{aligned} G(u,t) &= \frac{e^{-\lambda t}\frac{u}{1-u}}{1+\frac{u}{1-u}e^{-\lambda t}} = \frac{ue^{-\lambda t}}{1-u\left(1-e^{-\lambda t}\right)} = \\ &= ue^{-\lambda t}\sum_{k=0}^{\infty} u^k\left(1-e^{-\lambda t}\right)^k = \\ &= \sum_{k=1}^{\infty} u^k e^{-\lambda t}\left(1-e^{-\lambda t}\right)^{k-1} \end{aligned} \tag{2.34}$$

and this confirms (2.27).

From the probability generating function, by means of straightforward calculation, we can obtain that

$$\begin{aligned} \mathbb{E}[\hat{N}(t)] &= e^{\lambda t} \\ \mathbb{V}\text{ar}[\hat{N}(t)] &= e^{\lambda t}\left(e^{\lambda t}-1\right) \end{aligned} \tag{2.35}$$

Thus a linear birth process has a mean and variance increasing exponentially. However, given that

$$\sum_{k=1}^{\infty} \frac{1}{\lambda k} = \infty \tag{2.36}$$

$\Pr\{\hat{N}(t) < \infty\} = 1$ for $t > 0$, that is the Yule–Furry process does not explode. Moreover,

$$\mathbb{E}[\hat{N}(t) \mid \hat{N}(t_{i-1})] = \hat{N}(t_{i-1})e^{\lambda t} \tag{2.37}$$

We note that the explosion of the process $\hat{N}(t)$, $t > 0$ in a finite time is avoided if and only if

$$Z_\infty = T_1 + T_2 + \ldots + T_\infty = \infty \tag{2.38}$$

where T_j, $j \geq 1$, are the intertimes between successive jumps. In general, for the non-linear process, it holds true that

$$\begin{aligned}
\mathbb{E}\left[e^{-Z_\infty}\right] = \mathbb{E}\left[\exp\left\{-\sum_{j=1}^{\infty} T_j\right\}\right] &= \\
= \lim_{n\to\infty} \prod_{j=1}^{n} \mathbb{E}[e^{-T_j}] &= \\
= \lim_{n\to\infty} \prod_{j=1}^{n} \frac{\lambda_j}{1+\lambda_j} &= \\
= \prod_{j=1}^{\infty} \frac{1}{1+\frac{1}{\lambda_j}} &= \\
= \frac{1}{1+\sum_{j=1}^{\infty}\frac{1}{\lambda_j}+\dots}
\end{aligned} \tag{2.39}$$

Hence, if $\sum_{j=1}^{\infty} \frac{1}{\lambda_j} = \infty$, we have that $e^{-Z_\infty} = 0$ almost surely, i.e. $Z_\infty = \infty$.

2.3 Estimation of the Pure-birth Process

The estimation problem for the simple (linear) case is illustrated both in the continuous and discrete time sampling schemes.

2.3.1 Estimators

Continuous time sampling scheme. Theorem 3.1 in Keiding (1974) reports that the likelihood function for the pure-birth process is

$$L(\lambda) = (X_t - 1)_{(B_t)} \lambda^{B_t} e^{-\lambda S_t} \tag{2.40}$$

where X_t is the population size at time t and $S_t = \int_0^\infty X_u du =$ is the total time lived by the population X up to time t; $(a)_n$ is the Pochammer symbol (falling factorial), i.e. $a(a-1)...(a-n+1)$ and B_t is the number of offsprings occured.

Theorem 3.1 in Keiding (1974) also states that, under *continuous observation* of the process, the statistic (X_t, S_t), is a *sufficient statistic* and the maximum likelihood estimator $\tilde{\lambda}$ is:

$$\tilde{\lambda} = \frac{(X_t - m)}{S_t} = \frac{B_t}{S_t} \tag{2.41}$$

where m is the process initial value (number of progenitors), i.e. $X_0 = m$, and B_t is the number of offsprings.

Keiding (1974) derives asymptotic convergence results for the occurrence-exposure maximum likelihood estimator $\hat{\lambda} = \frac{B_t}{S_t}$, such that $\hat{\lambda}$ is strongly consistent for λ and it converges to a Gaussian distribution (see Theor. 3.5 in Keding, 1974) for $m \to \infty$ and fixed t, and it converges to a t-Student distribution (see Theor. 3.4 in Keding, 1974). More precisely, for fixed m, as $t \to \infty$:

$$\sqrt{\lambda S_t} \cdot \left(\frac{\hat{\lambda}}{\lambda} - 1 \right) \to N(0,1) \tag{2.42}$$

$$\sqrt{m e^{\lambda t}} \cdot \left(\frac{\hat{\lambda}}{\lambda} - 1 \right) \to t(2m) \tag{2.43}$$

For fixed t, as $m \to \infty$, then (2.42) holds and (2.43) is asymptotically $N\left(0, \frac{1}{1-e^{-\lambda t}}\right)$.

Moreover, Beyer, Keiding and Simonsen (1976) discuss asymptotic approximations for the moments of the estimators based on the two limiting distributions (assumed as an approximation to the true distribution), so that $\mathbb{E}[\hat{\lambda}] = \lambda$ and, respectively:

$$\mathbb{V}\left[\frac{\hat{\lambda}}{\lambda}\right] = \frac{1}{m(e^{\lambda t} - 1)} \tag{2.44}$$

$$\mathbb{V}\left[\frac{\hat{\lambda}}{\lambda}\right] = \left[(m-1)(e^{\lambda t} - 1)\right]^{-1} \tag{2.45}$$

Finally, Theorem 4.1 in Keiding (1976) provides, when the number of progenitors $m \to \infty$, more refined *asymptotic approximations* to the expectation, variance and skewness of the *relative bias* $\frac{\hat{\lambda}}{\lambda}$. For example, for the expected bias $\mathbb{E}\left[\frac{\hat{\lambda}}{\lambda}\right]$, where $\gamma = \lambda t$,

$$\mathbb{E}\left[\frac{\hat{\lambda}}{\lambda}\right] = 1 - \frac{\gamma e^{\gamma} - e^{\gamma} + 1}{m(e^{\gamma} - 1)^2} + O\left(\frac{1}{m^2}\right) \tag{2.46}$$

Discrete sampling scheme. Kendall (1949) developed the maximum likelihood inference of the simple (linear) pure-birth process under a discrete sampling setting, i.e. assuming n observations $\underline{\boldsymbol{x}} \equiv (x_0, x_1, ..., x_n)$ sampled over a time period $[0, T)$ at *equidistant* time points t_i, $i = 0, 1, 2, ..., n$ at a time step $\Delta t = \frac{T}{n}$, such that $t_0 = 0$, $t_n = T$ and $X(t_i) = x_i$.

The likelihood function was found by Kendall (1949) and is (see also Keiding (1974)):

$$L(\lambda; \boldsymbol{x}) = \\ = \prod_{i=1}^{n} \binom{x_i - 1}{x_i - x_{i-1}} \cdot \exp\left(-\lambda \Delta t \sum_{i=1}^{n} x_{i-1}\right) (1 - \exp(-\lambda \Delta t))^{x_n - x_0} \tag{2.47}$$

and the maximum likelihood estimator $\hat{\lambda}$ is (see Kendall, 1949, eq. 78)

$$\hat{\lambda} = \frac{1}{\Delta t} \log\left(\frac{x_1 + ... + x_n}{x_0 + ... + x_{n-1}}\right) \tag{2.48}$$

Theorem 5.2 in Keiding (1974) states the *strong consistency* of the estimator $\hat{\lambda}_n$, i.e. as the sample size $n \to \infty$, $\hat{\lambda}_n \to \lambda$.

Moreover, as $n \to \infty$

$$\Delta t \sqrt{(X_0 + ... + X_n)}(\hat{\lambda} - \lambda) \to N(0, 1 - e^{-\lambda \Delta t}) \tag{2.49}$$

$$\Delta t \frac{\sqrt{(m e^{\lambda(n+1)\Delta t} - 1) e^{\lambda \Delta t}}}{e^{\lambda \Delta t} - 1} (\hat{\lambda} - \lambda) \to t(2m) \tag{2.50}$$

If $n = 1$ and only X_t is observed, then $\hat{\lambda} = \frac{1}{t}\ln(\frac{X_t}{m})$ is the MLE and

$$t(\hat{\lambda} - \lambda) \to \ln(W) \quad a.s. \tag{2.51}$$

with W distributed as Gamma$(m, \frac{1}{m})$,

$$\mathbb{E}[\ln(W)] = \Psi(m) - \ln(m) \tag{2.52}$$

$$\mathbb{V}ar[\ln(W)] = \Psi'(m) \tag{2.53}$$

where Ψ is the digamma function $\Psi(x) = \frac{d\ln(\Gamma(x))}{dx}$ (Keiding, 1974, theor. 5.3).

3

Death Processes

CONTENTS

First the nonlinear case of the death process is introduced, for which the state distribution is obtained for a population of n_0 individuals.

Then the linear case is presented, for which a Binomial size distribution holds.

Moreover, the *sublinear* case – when the death rate decreases with the number of deaths occurred – is introduced, and its size distribution is provided.

Finally, for the continuous-time sampling scheme the process maximum likelihood estimator is stated as an occurrence-exposure ratio (see Keiding, 1974).

3.1 The Nonlinear Case

We consider a population with n_0 elements at time $t = 0$. Let $\mathcal{M}(t)$ be the number of elements of the population at time $t > 0$ and we denote by $\mathcal{M}(t, t + dt]$ the r.v. representing the decrease of the population in the time interval $(t,\ t + dt]$.

DOI: 10.1201/9781003398011-3

We suppose that the population can only decrease according to the following probabilistic rules

1. $\Pr\{\mathcal{M}(0) = n_0\} = 1$;
2. $\Pr\{\mathcal{M}(t, t+dt] = -1 \mid \mathcal{M}(t) = k\} = \mu_k dt + o(dt)$, $\mu_k > 0$;
3. $\Pr\{\mathcal{M}(t, t+dt] = 0 \mid \mathcal{M}(t) = k\} = 1 - \mu_k dt + o(dt)$;
4. $\Pr\{\mathcal{M}(t, t+dt] < -1 \mid \mathcal{M}(t) = k\} = o(dt)$,

The last property excludes simultaneous deaths.

The distribution

$$p_k(t) = \Pr\{\mathcal{M}(t) = k \mid \mathcal{M}(0) = n_0\} \tag{3.1}$$

is obtained by solving the following difference-differential equations

$$\begin{cases} \frac{dp_k(t)}{dt} = -\mu_k p_k(t) + \mu_{k+1} p_{k+1}(t), & 1 \leq k < n_0, \\ \frac{dp_{n_0}(t)}{dt} = -\mu_{n_0} p_{n_0}(t), & k = n_0, \quad \text{(initial condition)} \\ \frac{dp_0(t)}{dt} = \mu_1 p_1(t), & k = 0. \quad \text{(extinction)} \end{cases} \tag{3.2}$$

The equations (3.2) are obtained by considering that the probability that at time $t + dt$ there are k survivors is the probability that at time t either there are k survivors and in $(t, t+dt]$ no death is recorded, or there are $k+1$ survivors and one of them disappears during the same time interval, i.e.

$$p_k(t + dt) = p_k(t)(1 - \mu_k dt) + p_{k+1}(t)\mu_{k+1} dt + o(dt). \tag{3.3}$$

The second of equations (3.2) comes from

$$p_{n_0}(t + dt) = p_{n_0}(t)(1 - \mu_{n_0} dt) + o(dt) \tag{3.4}$$

and the last of equations (3.2) comes from

$$p_0(t + dt) = p_0(t) + p_1(t)\mu_1 dt + o(dt). \tag{3.5}$$

A relevant feature for the death process is the extinction probability $p_0(t)$. Note that the second and third of equations (3.2) are obtained from the first one for $k = n_0$ and $k = 0$, respectively.

The second equation of (3.2) comes from the first one with $\mu_{n_0+1} = 0$ because the population cannot increase in a pure-death process.

Theorem 3.1: The probability distribution emerging from (3.2), has the form

$$p_k(t) = \begin{cases} e^{-\mu_{n_0} t}, & k = n_0, \\ \prod_{j=k+1}^{n_0} \mu_j \sum\limits_{m=k}^{n_0} \frac{e^{-\mu_m t}}{\prod_{h=k,h\neq m}^{n_0} (\mu_h - \mu_m)}, & 1 \leq k < n_0, \\ 1 - \sum\limits_{m=1}^{n_0} \prod_{h=1,h\neq m}^{n_0} \left(\frac{\mu_h}{\mu_h - \mu_m}\right) e^{-\mu_m t}, & k = 0. \end{cases} \tag{3.6}$$

Proof. The first result is straightforward. As far as the second probability of (3.6) is concerned, we restrict ourselves to some specific cases. Equation (3.2), for $k = n_0 - 1$ becomes

$$\begin{aligned} \frac{dp_{n_0-1}(t)}{dt} &= \\ &= -\mu_{n_0-1} p_{n_0-1}(t) + \mu_{n_0} p_{n_0}(t) \\ &= -\mu_{n_0-1} p_{n_0-1}(t) + \mu_{n_0} e^{-\mu_{n_0} t}, \end{aligned} \tag{3.7}$$

and its solution is

$$\begin{aligned} p_{n_0-1}(t) &= \\ &= e^{-t\mu_{n_0-1}} \left\{ \int_0^t \mu_{n_0} e^{-s\mu_{n_0}} e^{s\mu_{n_0-1}} + \text{cost} \right\} = \\ &= \mu_{n_0} e^{-t\mu_{n_0-1}} \left\{ \frac{1 - e^{-(\mu_{n_0} - \mu_{n_0-1})t}}{\mu_{n_0} - \mu_{n_0-1}} \right\} = \\ &= \mu_{n_0} \left[\frac{e^{-t\mu_{n_0-1}}}{\mu_{n_0} - \mu_{n_0-1}} + \frac{e^{-t\mu_{n_0}}}{\mu_{n_0-1} - \mu_{n_0}} \right], \end{aligned} \tag{3.8}$$

and is equal to (3.6) for $k = n_0 - 1$.

The recursion method permits us to write

$$\frac{dp_{n_0-2}(t)}{dt} = \\ = -\mu_{n_0-2}p_{n_0-2}(t) + \\ +\mu_{n_0-1}\mu_{n_0}\left[\frac{e^{-t\mu_{n_0-1}}}{\mu_{n_0}-\mu_{n_0-1}} + \frac{e^{-t\mu_{n_0}}}{\mu_{n_0-1}-\mu_{n_0}}\right], \tag{3.9}$$

with initial condition $p_{n_0-2}(0) = 0$.

The solution of (3.9) reads

$$\begin{aligned}
p_{n_0-2}(t) &= \\
= e^{-t\mu_{n_0-2}}\left\{\int_0^t \mu_{n_0-1}\mu_{n_0}\left[\frac{e^{-s\mu_{n_0-1}}}{\mu_{n_0}-\mu_{n_0-1}} + \frac{e^{-s\mu_{n_0}}}{\mu_{n_0-1}-\mu_{n_0}}\right] e^{s\mu_{n_0-2}}ds\right\} &= \\
= \mu_{n_0}\mu_{n_0-1}e^{-t\mu_{n_0-2}}\left[\frac{1-e^{-(\mu_{n_0-1}-\mu_{n_0-2})t}}{(\mu_{n_0}-\mu_{n_0-1})(\mu_{n_0}-\mu_{n_0-2})} + \right. \\
\left. - \frac{1-e^{-(\mu_{n_0}-\mu_{n_0-2})t}}{(\mu_{n_0}-\mu_{n_0-1})(\mu_{n_0}-\mu_{n_0-2})}\right] &= \\
= \mu_{n_0}\mu_{n_0-1}\left[\frac{e^{-t\mu_{n_0-2}}}{(\mu_{n_0}-\mu_{n_0-1})(\mu_{n_0-1}-\mu_{n_0-2})} + \right. \\
- \frac{e^{-t\mu_{n_0-2}}}{(\mu_{n_0}-\mu_{n_0-1})(\mu_{n_0}-\mu_{n_0-2})} + \\
\left. + \frac{e^{-t\mu_{n_0-1}}}{(\mu_{n_0}-\mu_{n_0-1})(\mu_{n_0-2}-\mu_{n_0-1})} + \frac{e^{-t\mu_{n_0}}}{(\mu_{n_0-1}-\mu_{n_0})(\mu_{n_0-2}-\mu_{n_0})}\right] &= \\
= \mu_{n_0}\mu_{n_0-1}\left[\frac{e^{-t\mu_{n_0}}}{(\mu_{n_0-1}-\mu_{n_0})(\mu_{n_0-2}-\mu_{n_0})} + \right. \\
+ \frac{e^{-t\mu_{n_0-1}}}{(\mu_{n_0}-\mu_{n_0-1})(\mu_{n_0-2}-\mu_{n_0-1})} + \\
\left. + \frac{e^{-t\mu_{n_0-2}}}{(\mu_{n_0}-\mu_{n_0-2})(\mu_{n_0-1}-\mu_{n_0-2})}\right] &=
\end{aligned} \tag{3.10}$$

Note that (3.6) for $k = n_0 - 2$ is equal to (3.10). The general case can be obtained in a similar way but the details are omitted (see Orsingher, Polito and Sakhno, 2010).

3.2 The Linear Case

The special case of the linear death process is retrieved for $\mu_k = k\mu$, so the mortality rate is proportional to the current size of the population.

By observing that

$$\begin{aligned} \prod_{h=k,h\neq m}^{n_0} (\mu_h - \mu_m) &= \\ = \mu^{n_0-k} \prod_{h=k,h\neq m}^{n_0} (h-m) &= \\ = \mu^{n_0-k} \cdot [(n_0 - m)(n_0 - 1 - m) \cdots 1(-1)(-2) \cdots (k-m)] &= \\ = \mu^{n_0-k}(-1)^{m-k} \cdot [(n_0 - 1) \cdot (n_0 - 1 - m) \cdots 1][1 \cdot 2 \cdots (m-k)] &= \\ = \mu^{n_0-k}(-1)^{m-k}(m-k)!(n_0-m)! & \end{aligned} \tag{3.11}$$

and that

$$\begin{aligned} \prod_{h=k+1}^{n_0} \mu_h = \prod_{h=k+1}^{n_0} \mu_h \cdot \frac{\prod_{h=1}^{k} \mu_h}{\prod_{h=1}^{k} \mu_h} = \frac{\prod_{h=1}^{n_0} \mu_h}{\prod_{h=1}^{k} \mu_h} &= \\ = \mu^{n_0-k} \frac{n_0!}{k!} & \end{aligned} \tag{3.12}$$

formula (3.6) entails that

$$\begin{aligned} p_k(t) &= \frac{n_0!}{k!} \sum_{m=k}^{n_0} \frac{e^{-\mu m t}(-1)^{m-k}}{(m-k)!(n_0-m)!} = \\ &= \binom{n_0}{k} \sum_{m=k}^{n_0} \binom{n_0-k}{m-k} (-1)^{m-k} e^{-\mu m t} = \\ &= \binom{n_0}{k} \sum_{r=0}^{n_0-k} \binom{n_0-k}{r} (-1)^r e^{-\mu(k+r)t} = \\ &= \binom{n_0}{k} e^{-\mu k t} \left(1-e^{-\mu t}\right)^{n_0-k}, \quad 0 \le k \le n_0 \end{aligned} \tag{3.13}$$

so the distribution of the linear death process is Binomial (see Fig. 3.1) with parameters n_0 and $e^{-\mu t}$ (see Orsingher, Polito and Sakhno, 2010).

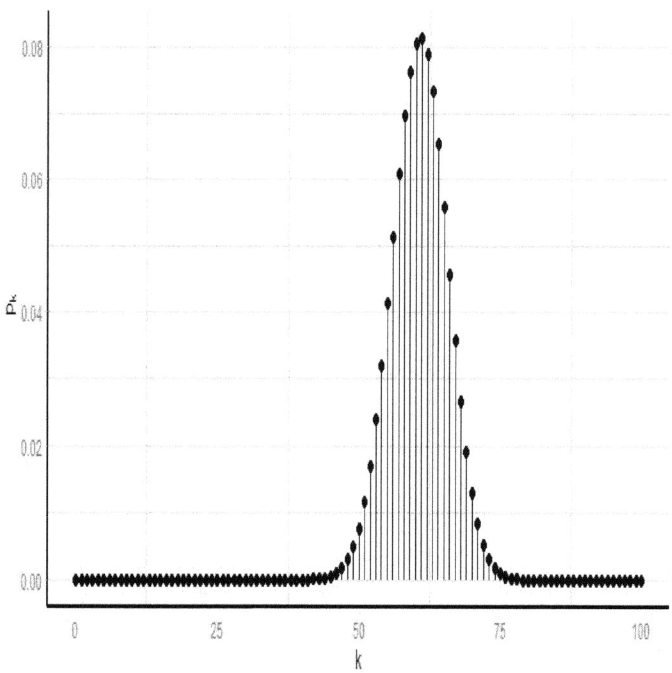

FIGURE 3.1
State probabilities $\{p_k(t),\, k \ge 1\}$ of a pure-death process (linear case) with initial population size $n_0 = 100$, after $t = 1$ time unit ($\mu = 0.1$). Binomial$(n_0, e^{-\mu t})$.

We have, clearly:

$$\mathbb{E}[M(t)] = n_0 e^{-\mu t} \tag{3.14}$$

$$\mathbb{V}ar[M(t)] = n_0 e^{-\mu t}(1 - e^{-\mu t}) \tag{3.15}$$

$$\mathbb{E}[M(t) \mid M(t-1)] = M(t-1)e^{-\mu t} \tag{3.16}$$

and the probability of extinction is

$$\Pr\{M(t) = 0 \mid M(0) = n_0\} = (1 - e^{-\mu t})^{n_0} \tag{3.17}$$

3.3 The Sublinear Case

As shown in Orsingher, Polito and Sakhno (2010), in this case the infinitesimal death probabilities have the form

$$\Pr\{\mathcal{M}(t, t+dt) = -1 \mid \mathcal{M}(t) = k\} = \mu(n_0+1-k)dt+o(dt) \tag{3.18}$$

where the death rate does not depend on the d number k of survivors, but on the number $n_0 - (k-1)$ of deaths occurred in the time interval $(0, t)$, so that for $k = n_0$ the probability (3.18) reduces to is $\mu dt + o(dt)$.

The difference-differential equations governing the survival probabilities $p_k(t)$ are

$$\frac{dp_k(t)}{dt} = -\mu(n_0 - k + 1)p_k(t) + \mu(n_0 - k)p_{k+1}(t) \tag{3.19}$$

subject to

$$\begin{cases} \frac{dp_{n_0}(t)}{dt} = -\mu p_{n_0}(t) \\ \frac{dp_0(t)}{dt} = \mu n_0 p_1(t) \end{cases} \tag{3.20}$$

and

$$p_k(0) = \begin{cases} 1 & k = n_0 \\ 0 & 0 \le k < n_0 \end{cases} \tag{3.21}$$

with solution given in Orsingher, Polito and Sakhno (2010)

$$p_k(t) = \sum_{h=0}^{n_0-k} \binom{n_0-k}{h} (-1)^h e^{-(h+1)\mu t} \quad 1 \le k \le n_0 \tag{3.22}$$

and extinction probability

$$p_0(t) = \sum_{h=0}^{n_0} \binom{n_0}{h} (-1)^h e^{-h\mu t}. \tag{3.23}$$

3.4 Estimation of the Pure-death Process

This section presents the point estimation of the death rate parameter in the simple (linear) pure-death process in the continuous time scheme.

3.4.1 Estimators

Continuous time sampling scheme.

Beyer, Keiding and Simonsen (1976) state that, under *continuous observation* of the process, the statistic (D_t, S_t), is a *sufficient statistic* and the maximum likelihood estimator $\tilde{\lambda}$ is:

$$\hat{\mu} = \frac{D_t}{S_t} \tag{3.24}$$

where D_t is the number of deaths by time t and $S_t = \int_0^\infty X_u du =$ is the total time lived by the population X up to time t.

See also §4.4, where the death process is obtained as $\lambda = 0$.

4

The Birth-death Process

CONTENTS

The birth-death process is introduced starting from its nonlinear version by stating the evolution rules.

Then the linear case is presented beginning with an ample discussion of its size distribution both in the case of one single progenitor and of $n_0 > 1$ progenitors.

Moreover, the process feature of a random walk in continuous-time is hinted at stating the Cauchy problem of its sojourn times and an example of the linear growth with immigration is discussed.

The p.g.f. is derived as a solution of a linear PDE and an ample treatment of the extinction probability is provided as a solution of the famous Riccati equation.

Finally, classical results of Keiding (1975) on the process of maximum likelihood inference are presented both w.r.t. the discrete-sampling scheme, for which Kendall's estimator holds, and the continuous observation scheme, where subject to supplementary observational hypotheses (see Keiding, 1975), an exposure-occurrence ratio provides the maximum likelihood estimator for the birth, death and Malthusian rates.

An application to the population of bears in the Trentino region is provided, based on annual *census data* and by applying estimators of an approximating branching (Galton-Watson) process.

DOI: 10.1201/9781003398011-4

4.1 The Nonlinear Case

We consider a process where birth and death are admitted so that the population $\mathcal{N}(t)$ can increase and decrease as time passes. We assume that

$$\begin{cases} \Pr\{(\mathcal{N}(t,t+dt)=1 \mid \mathcal{N}(t)=k\} = \lambda_k dt + o(dt) \\ \Pr\{(\mathcal{N}(t,t+dt)=-1 \mid \mathcal{N}(t)=k\} = \mu_k dt + o(dt) \\ \Pr\{(\mathcal{N}(t,t+dt)=0 \mid \mathcal{N}(t)=k\} = 1-(\lambda_k+\mu_k)dt + o(dt) \end{cases} \tag{4.1}$$

where μ_k and λ_k are the death and birth rates.

Furthermore, we assume also that

$$\Pr\{\mathcal{N}(t,t+dt)=\pm m \mid \mathcal{N}(t)=k\} = o(dt), \quad 2 \leq m \leq k \tag{4.2}$$

i.e. we suppose that clusters of deaths and births occur with a negligible probability. The state probabilities

$$p_k(t) = \Pr\{\mathcal{N}(t)=k \mid \mathcal{N}(0)=1\}, k \geq 0 \tag{4.3}$$

solve the system of equations

$$\frac{dp_k(t)}{dt} = -(\lambda_k+\mu_k)p_k(t) + \lambda_{k-1}p_{k-1}(t) + \mu_{k+1}p_{k+1}(t), \qquad k \geq 0 \tag{4.4}$$

with $p_{-1}(t)=0$ and under initial conditions

$$p_k(0) = \begin{cases} 1 & k=1 \qquad \text{(one initial progenitor)} \\ 0 & k \neq 1. \end{cases} \tag{4.5}$$

Equation (4.4) can be obtained in the following manner:

$$\begin{aligned} &p_k(t+dt) = \\ &= p_k(t)(1-\lambda_k dt)(1-\mu_k dt)+ \\ &+p_{k-1}(t)\lambda_{k-1}dt(1-\mu_{k-1}dt) + p_{k+1}(t)(1-\lambda_{k+1}dt)\mu_{k+1}dt+ \\ &+p_k(t)\lambda_k dt \mu_k dt = \\ &= p_k(t) - \lambda_k p_k(t)dt - \mu_k p_k(t)dt+ \\ &+ p_{k-1}(t)\lambda_{k-1}dt + p_{k+1}(t)\mu_{k+1}dt \end{aligned} \tag{4.6}$$

By expanding $p_k(t+dt)$, dividing both members by dt we get (4.4) in the limit.

Remark 4.1.
If the Poisson rates are constant, $\lambda_k = \lambda$, $\mu_k = \mu$, the birth-death process becomes the immigration-emigration process.

Remark 4.2.
If the Poisson rates are time dependent, that is $\lambda_k = \lambda_k(t)$ and $\mu_k = \mu_k(t)$, the resulting process is referred to as *general* birth-death process and was studied by Kendall (1948); see also remark 2.1.

4.2 The Linear Case

For the *linear* birth-and-death process, we obtain:

$$\frac{dp_k(t)}{dt} = -(\lambda+\mu)kp_k(t) + \lambda(k-1)p_{k-1}(t) + \mu(k+1)p_{k+1}(t) \tag{4.7}$$

Theorem 4.1. *The exact distribution* $p_k(t) = Pr\{\mathcal{N}(t) = k \mid \mathcal{N}(0) = 1\}$, $k \geq 1$[1], *is*

$$p_k(t) = (\lambda-\mu)^2 e^{-(\lambda-\mu)t} \frac{[\lambda(1-e^{-(\lambda-\mu)t})]^{k-1}}{(\lambda-\mu e^{-(\lambda-\mu)t})^{k+1}}, \qquad k \geq 1, \mu \neq \lambda \tag{4.8}$$

[1]See also Bailey (1964), ch. 8, eq. (8.46).

Proof. Let the probability generating function (see equation (4.21) of the linear birth-death process be expressed as

$$\begin{aligned}G(u,t) &= \frac{\mu(1-u)+(\lambda u-\mu)e^{-(\lambda-\mu)t}}{\lambda(1-u)+(\lambda u-\mu)e^{-(\lambda-\mu)t}} = \\ &= \frac{\mu-\mu e^{-(\lambda-\mu)t}-(\mu-\lambda e^{-(\lambda-\mu)t})u}{\lambda-\mu e^{-(\lambda-\mu)t}-(\lambda-\lambda e^{-(\lambda-\mu)t})u} = \\ &= \frac{[\mu-\mu e^{-(\lambda-\mu)t}][1-\frac{(\mu-\lambda e^{-(\lambda-\mu)t})u}{\mu-\mu e^{-(\lambda-\mu)t}}]}{\lambda-\mu e^{-(\lambda-\mu)t}[1-\frac{(\lambda-\lambda e^{-(\lambda-\mu)t})u}{\lambda-\mu e^{-(\lambda-\mu)t}}]} = \\ &= C\left[\frac{1-Au}{1-Bu}\right]\end{aligned} \tag{4.9}$$

where

$$\begin{cases} A = \frac{\mu-\lambda e^{-(\lambda-\mu)t}}{\mu(1-e^{-(\lambda-\mu)t})} \\ B = \frac{\lambda(1-e^{-(\lambda-\mu)t})}{\lambda-\mu e^{-(\lambda-\mu)t}} \\ C = \frac{\mu(1-e^{-(\lambda-\mu)t})}{\lambda-\mu e^{-(\lambda-\mu)t}} \end{cases} \tag{4.10}$$

If $|Bu| < 1$, we can rewrite

$$\begin{aligned}G(u,t) = C(1-Au)\left(\frac{1}{1-Bu}\right) &= C(1-Au)\sum_{k=0}^{\infty}(Bu)^k = \\ &= C\left[\sum_{k=0}^{\infty}(Bu)^k - \sum_{k=0}^{\infty}u^{k+1}AB^k\right] = \\ &= C\left[1+\sum_{k=1}^{\infty}(Bu)^k - \sum_{k=1}^{\infty}u^k AB^{k-1}\right] = \\ &= C\left[1+\sum_{k=1}^{\infty}(Bu)^k\left(1-\frac{A}{B}\right)\right] = \\ &= \sum_{k=0}^{\infty}u^k p_k(t)\end{aligned} \tag{4.11}$$

where

$$\begin{cases} p_0(t) = C \\ p_k(t) = C\left(1 - \frac{A}{B}\right) B^k, & k \geq 1 \end{cases} \tag{4.12}$$

We now evaluate

$$B - A = \lambda\left(\frac{1 - e^{-(\lambda-\mu)t}}{\lambda - \mu e^{-(\lambda-\mu)t}}\right) - \frac{\mu - \lambda e^{-(\lambda-\mu)t}}{\mu(1 - e^{-(\lambda-\mu)t})} =$$

$$= \frac{\lambda\mu[1 - e^{-(\lambda-\mu)t}]^2 - [\lambda\mu - \lambda^2 e^{-(\lambda-\mu)t} - \mu^2 e^{-(\lambda-\mu)t} + \lambda\mu e^{-2(\lambda-\mu)t}]}{[\lambda - \mu e^{-(\lambda-\mu)t}]\mu[1 - e^{-(\lambda-\mu)t}]} =$$

$$= \frac{(\lambda - \mu)^2 e^{-(\lambda-\mu)t}}{\mu(\lambda - \mu e^{-(\lambda-\mu)t})(1 - e^{-(\lambda-\mu)t})} \tag{4.13}$$

Therefore

$$\begin{aligned} p_k(t) &= \\ &= \frac{\mu(1 - e^{-(\lambda-\mu)t})}{\lambda - \mu e^{-(\lambda-\mu)t}} \cdot \frac{(\lambda - \mu)^2 e^{-(\lambda-\mu)t}}{\mu(\lambda - \mu e^{-(\lambda-\mu)t})(1 - e^{-(\lambda-\mu)t})} \times \\ &\times \left[\frac{\lambda(1 - e^{-(\lambda-\mu)t})}{\lambda - \mu e^{-(\lambda-\mu)t}}\right]^{k-1} = \\ &= (\lambda - \mu)^2 e^{-(\lambda-\mu)t} \frac{[\lambda(1 - e^{-(\lambda-\mu)t})]^{k-1}}{[\lambda - \mu e^{-(\lambda-\mu)t}]^{k+1}} \end{aligned} \tag{4.14}$$

□

For $\lambda = \mu$ simplifies into (see Fig. 4.1)

$$p_k(t) = \frac{(\lambda t)^{k-1}}{(1 + \lambda t)^{k+1}}, k \geq 1 \tag{4.15}$$

The extinction probabilities are (see Fig. 4.2)

$$p_0(t) = \begin{cases} \frac{\mu t}{1+\mu t} & \lambda = \mu \\ \frac{\mu - \mu e^{-(\lambda-\mu)t}}{\lambda - \mu e^{-(\lambda-\mu)t}} & \lambda \neq \mu \end{cases} \tag{4.16}$$

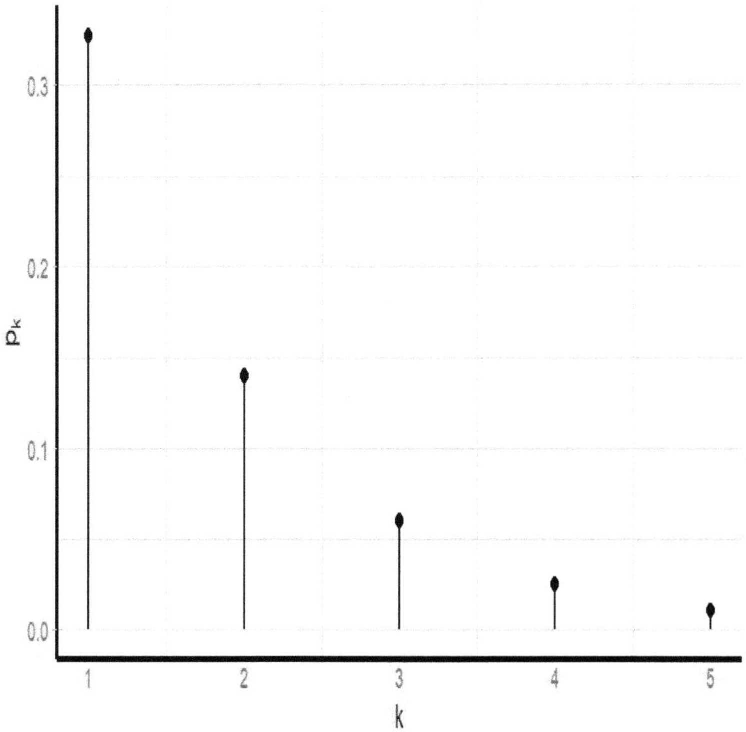

FIGURE 4.1
Distribution $\{p_k(t),\, k \geq 1\}$, for $\lambda = \mu = 0.75$; $t = 1$.

The probability generating function $G(u,t) = \mathbb{E}[u^{\mathcal{N}(t)}]$ of $\mathcal{N}(t)$ satisfies in the linear case the partial differential equation

$$\begin{aligned}\frac{\partial G(u,t)}{\partial t} &= \\ &= \left[\lambda u^2 - (\lambda+\mu)u + \mu\right] \frac{\partial G(u,t)}{\partial u} = \\ &= (\lambda u - \mu)(u-1)\frac{\partial G(u,t)}{\partial u},\ |u| \leq 1, t > 0\end{aligned} \tag{4.17}$$

with the initial condition $G(u,0) = u$ (with a single progenitor at time $t = 0$).

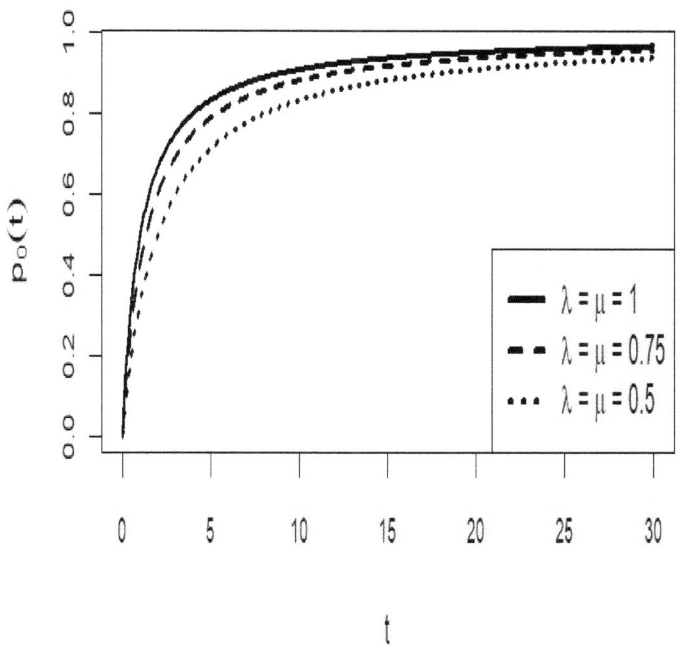

FIGURE 4.2
Extinction probabilities $p_0(t)$, for $\lambda = \mu = 1$, $\lambda = \mu = 0.75$ and $\lambda = \mu = 0.5$.

The general form of the solution of equation (4.17) is

$$G(u,t) = f\left(e^{-(\lambda-\mu)t}\frac{\lambda u - \mu}{1-u}\right) \tag{4.18}$$

and this can be checked by substituting equation (4.18) into equation (4.17) and assuming that $f \in \mathbb{C}^1$. The argument of equation (4.18) is obtained by the method of auxiliary functions. By imposing the initial condition

$$G(u,0) = \sum_{k=0}^{\infty} u^k p_k(0) = u = f\left(\frac{\lambda u - \mu}{1-u}\right) \tag{4.19}$$

we obtain that

$$f(u) = \frac{\mu + u}{\lambda + u} \tag{4.20}$$

because if we set $\frac{\lambda u - \mu}{1-u} = v$, we readily arrive at (4.20).

Remark 4.3. Linear growth with immigration
If the birth rate increases linearly with the size k and benefits of an external contribution c such that $\lambda_k = \lambda k + c$, while the death rate is $\mu_k = \mu k$, then the process is referred to as *linear* birth-death process *with immigration* and c is the (constant) immigration rate.

The p.g.f. for $\lambda \neq \mu$ is therefore

$$G(u,t) = \begin{cases} \frac{\mu + \frac{\lambda u - \mu}{1-u} e^{-(\lambda-\mu)t}}{\lambda + \frac{\lambda u - \mu}{1-u} e^{-(\lambda-\mu)t}} = \frac{\mu(1-u) + (\lambda u - \mu) e^{-(\lambda-\mu)t}}{\lambda(1-u) + (\lambda u - \mu) e^{-(\lambda-\mu)t}} & u > 0 \\ p_0(t) = \frac{\mu - \mu e^{-(\lambda-\mu)t}}{\lambda - \mu e^{-(\lambda-\mu)t}}, & u = 0 \end{cases} \tag{4.21}$$

The limit of the extinction probability (4.21) for $\lambda \to \mu$ reads (see Fig. 4.2)

$$\lim_{\lambda\to\mu} \frac{\mu - \mu e^{-(\lambda-\mu)t}}{\lambda - \mu e^{-(\lambda-\mu)t}} = \frac{\mu t}{1 + \mu t}. \tag{4.22}$$

The distribution for $n_0 > 1$ progenitors. We consider a linear birth and death process $\{\mathfrak{N}(t), t \geq 0\}$ with $n_0 > 1$ initial progenitors. Each process develops independently, and thus we must have in mind the characteristics of a birth and death process with one initial progenitor. In the case of a single progenitor ($n_0 = 1$), the distribution (4.8) $\Pr\{N(t) = k \mid N(0) = 1\}$ (representing the number of progenies of a single progenitor) was found by Kendall (1948) to be *modified* Geometric (a Geometric distribution $\{p_k(t), k \geq 1\}$, extended to the extinction probability $p_0(t)$ (4.16)), i.e. ($\lambda \neq \mu$):

$$p_k(t) = p_0(t) \cdot 1_{\{k=0\}} + (1 - p_0(t)) \cdot (1 - \alpha(t))(\alpha(t))^{k-1} \cdot 1_{\{k \geq 1\}} \tag{4.23}$$

where $\alpha(t) = \frac{\lambda}{\mu} p_0(t)$.

More precisely, the distribution, at time t, of each progeny N_j

$$\Pr\{N_j(t) = k \mid N_j(0) = 1\} = \\ = \frac{[(\lambda-\mu)^2 e^{-(\lambda-\mu)t}(\lambda(1-e^{-(\lambda-\mu)t}))]^{k-1}}{[\lambda-\mu e^{-(\lambda-\mu)t}]^{k+1}}, \qquad k \geq 1 \tag{4.24}$$

can be expressed in terms of the extinction probability

$$\Pr\{N_j(t) = 0 \mid N_j(0) = 1\} = p_0(t) = \\ = \frac{\mu - \mu e^{-(\lambda-\mu)t}}{\lambda - \mu e^{-(\lambda-\mu)t}} \qquad \lambda \neq \mu \tag{4.25}$$

In fact, we have that

$$\Pr\{N_j(t) = k \mid N_j(0) = 1\} = \\ = (1-p_0(t))\,(1-\alpha(t))\,(\alpha(t))^{k-1} = \\ = (1-p_0(t))(1-\alpha(t))\alpha(t)^{k-1}, \qquad \text{with } \alpha(t) = \frac{\lambda}{\mu} p_0(t) \tag{4.26}$$

In order to check (4.26) we observe that

$$(1-p_0(t))(1-\alpha(t)) = \\ = \left[1 - \frac{\mu - \mu e^{-(\lambda-\mu)t}}{\lambda - \mu e^{-(\lambda-\mu)t}}\right] \cdot \left[1 - \frac{\lambda}{\mu}\frac{\mu - \mu e^{-(\lambda-\mu)t}}{\lambda - \mu e^{-(\lambda-\mu)t}}\right] = \\ = \left[1 - \frac{\mu - \mu e^{-(\lambda-\mu)t}}{\lambda - \mu e^{-(\lambda-\mu)t}}\right] \cdot \left[1 - \frac{\lambda - \lambda e^{-(\lambda-\mu)t}}{\lambda - \mu e^{-(\lambda-\mu)t}}\right] = \\ = \frac{(\lambda-\mu)}{\lambda - \mu e^{-(\lambda-\mu)t}} \cdot \frac{(\lambda-\mu)e^{-(\lambda-\mu)t}}{\lambda - \mu e^{-(\lambda-\mu)t}} = \\ = \left(\frac{\lambda-\mu}{\lambda - \mu e^{-(\lambda-\mu)t}}\right)^2 e^{-(\lambda-\mu)t} \tag{4.27}$$

In conclusion

$$(1-p_0(t))(1-\alpha(t)) = \frac{(\lambda-\mu)^2 e^{-(\lambda-\mu)t}}{(\lambda - \mu e^{-(\lambda-\mu)t})^2} \tag{4.28}$$

Thus the distribution (4.24) can thus be written as

$$\begin{aligned}
&\Pr\{N_j(t) = k \mid N_j(t) = 1\} = \\
&= \frac{(\lambda-\mu)^2 e^{-(\lambda-\mu)t}}{(\lambda-\mu e^{-(\lambda-\mu)t})^2} \cdot \left(\frac{\lambda - \lambda e^{-(\lambda-\mu)t}}{\lambda - \mu e^{-(\lambda-\mu)t}}\right)^{k-1} = \\
&= (1-p_0(t))(1-\alpha(t)) \cdot \alpha(t)^{k-1}, \text{ for } k \geq 1
\end{aligned} \tag{4.29}$$

The probability generating function of (4.23) becomes

$$\begin{aligned}
&\mathbb{E}\left[u^{X_j(t)} \mid X_j(0) = 1\right] = \\
&= p_0(t) + \sum_{k=1}^{\infty} u^k (1-p_0(t))(1-\alpha(t))\alpha(t)^{k-1} = \\
&= p_0(t) + u(1-p_0(t))(1-\alpha(t)) \cdot \sum_{k=1}^{\infty} (u\alpha(t))^{k-1} = \\
&= p_0(t) + \frac{u(1-p_0(t))(1-\alpha(t))}{1-u\alpha(t)} = \\
&= \frac{p_0(t) - up_0(t)\alpha(t) + u(1-\alpha(t) - p_0(t) + p_0(t)\alpha(t))}{1-u\alpha(t)} = \\
&= \frac{p_0(t) + u(1-p_0(t)-\alpha(t))}{1-u\alpha(t)}
\end{aligned} \tag{4.30}$$

In the case of n_0 progenitors (each generating a family $N_j(t)$), we have that the number of individuals at time t is

$$\mathfrak{N}(t) = \sum_{j=1}^{F(t)} N_j(t) \tag{4.31}$$

where $F(t)$ is the random number of families (progenies generated by a common progenitor) existing at time $t = 0$ and which survived at time t. The number of survivors at time t is independent of the $X_j(t)$, for all $j = 1, ..., F(t)$

$$\begin{aligned}
&\Pr\{\mathfrak{N}(t) = k \mid \mathfrak{N}(0) = n_0\} = \\
&= \Pr\left\{ \sum_{j=1}^{F(t)} N_j(t) = k \,\middle|\, \mathfrak{N}(0) = n_0 \right\} = \\
&= \mathbb{E}\left[\Pr\left\{ \sum_{j=1}^{F(t)} N_j(t) = k \right\} \,\middle|\, F(t) \right] = \\
&= \sum_{j=1}^{n_0} \Pr\left\{ \sum_{j=1}^{n} N_j(t) = k\} \cdot \Pr\{F(t) = n \right\}
\end{aligned} \tag{4.32}$$

Of course,

$$\Pr\{F(t) = n\} = \binom{n_0}{n} (1 - \alpha(t))^n \alpha(t)^{n_0 - n}, \qquad n = 0, 1, ..., n_0 \tag{4.33}$$

and $\Pr\{F(t) = 0\}$ represents the probability that all families generated by the initial n_0 progenitors have disappeared at time t.

In order to evaluate

$$\Pr\left\{ \sum_{j=1}^{n} N_j(t) = k \right\} \tag{4.34}$$

we resort to the probability generating function

$$\begin{aligned}
\mathbb{E}\left[u^{\sum_{j=1}^{n} N_j(t)} \right] &= \prod_{j=1}^{n} \left(\mathbb{E}\left[u^{N_j(t)} \right] \right) = \left(\mathbb{E}\left[u^{N_1(t)} \right] \right)^n = \\
&= \left(\frac{p_0(t) + u(1 - p_0(t) - \alpha(t))}{1 - u\alpha(t)} \right)^n = \\
&= (p_0(t) + u(1 - p_0(t) - \alpha(t)))^n \cdot \sum_{h=0}^{\infty} \binom{-n}{h} (-u\alpha(t))^h
\end{aligned} \tag{4.35}$$

Since

$$\begin{aligned}\binom{-n}{h} &= \frac{-n(-n-1)...(-n-h+1)}{h!} = \\ &= (-1)^h \frac{n(n+1)...(n+h-1)}{(n-1)!h!}(n-1)! = \\ &= (-1)^h \binom{n+h-1}{h} = (-1)^h \binom{n+h-1}{n-1}\end{aligned} \tag{4.36}$$

we obtain

$$\begin{aligned}&\mathbb{E}\left[u^{\sum_{j=1}^{n} N_j(t)}\right] = \left(\mathbb{E}\left[u^{N_j(t)}\right]\right)^n = \\ &= (p_0(t) + u(1-p_0(t)-\alpha(t)))^n \cdot \sum_{r=0}^{\infty} \binom{n+h-1}{n-1} \alpha(t)^h u^h = \\ &= \sum_{j=0}^{n} \binom{n}{j} u^j (1-p_0(t)-\alpha(t))^j p_0(t)^{n-j} \cdot \sum_{h=0}^{\infty} \binom{n+h-1}{n-1} \alpha(t)^h \cdot \\ &\cdot u^h = \\ &= \sum_{j=0}^{n} \sum_{h=0}^{\infty} \binom{n}{j} \binom{n+h-1}{n-1} (1-p_0(t)-\alpha(t))^j p_0(t)^{n-j} \alpha(t)^h u^{j+h} = \\ &= \sum_{j=0}^{n} \sum_{k=j}^{\infty} \binom{n}{j} \binom{n+k-j-1}{n-1} (1-p_0(t)-\alpha(t))^j p_0(t)^{n-j} \alpha(t)^{k-j} \cdot \\ &\cdot u^k = \\ &= \text{(isolating the coefficient of } u^k\text{)} \sum_{k=j}^{\infty} u^k \cdot \\ &\cdot \sum_{j=0}^{n} \binom{n}{j} \binom{n+k-1-j}{n-1} (1-p_0(t)-\alpha(t))^j p_0(t)^{n-j} \alpha(t)^{k-j}\end{aligned} \tag{4.37}$$

Thus

$$\Pr\left\{\sum_{j=1}^{n} N_j(t) = k\right\} = \text{ coefficient of } u^k =$$

$$= \sum_{j=0}^{n} \binom{n}{j} \binom{n+k-j-1}{n-1} p_0(t)^{n-j} \alpha(t)^{k-j} (1 - p_0(t) - \alpha(t))^j \tag{4.38}$$

where n is the number of surviving families which must be equal or less than n_0.

The expected number of *surviving* families $F(t)$, starting from n_0 progenitors, and its variance, are (see Tavare, 2018)

$$\begin{cases} \mathbb{E}[F(t) \mid N(0) = n_0] = n_0(1 - p_0(t)) \\ \mathbb{V}ar[F(t) \mid N(0) = n_0] = n_0 p_0(t)(1 - p_0(t)) \end{cases} \tag{4.39}$$

Remark. A result in Waugh (1958) shows by a probabilistic argument based on conditioning upon ultimate *extinction* that $p_k(t; \lambda, \mu) = \left(\frac{\mu}{\lambda}\right)^{k-n_0} \times p_k(t; \mu, \lambda)$, that is the state probabilities of a *supercritical* process (i.e. $\lambda > \mu$) can also be obtained by swapping the rate parameters and adjusting by the factor $\left(\frac{\mu}{\lambda}\right)^{k-n_0}$.

This property turns precious when computing the state probabilities and some parameters configuration (see Tavare, 2018) is numerically unstable.

4.3 The Extinction Probability and the Riccati Equation

The equation governing the extinction probability is

$$p_0(t) = \int_0^t e^{-\lambda s} \mu e^{-\mu s} ds + \int_0^t \lambda e^{-\lambda s} e^{-\mu s} p_0^2(t-s) ds. \tag{4.40}$$

Equation (4.40) is obtained by considering that

$$\int_0^t e^{-\lambda s}\mu e^{-\mu s} ds \tag{4.41}$$

represents the probability that in $(0, s)$ the initial particle will survive up to time s with probability $e^{-\mu s}$ (and no offspring was born, with probability $e^{-\lambda s}$), and that it will die during $[s, s+ds)$ with probability μds.

The second term of (4.40) considers that the initial progenitor with probability $e^{-\mu s}$ survives up to time s (and no birth is registered in $(0, s)$ with probability $e^{-\lambda s}$), an additional offspring is generated in $[s, s+ds)$ with probability λds, and both components of the population will die during the remaining time $[s, t]$, independently, with probability $p_0^2(t-s)$. Hence, by deriving with respect to t, we obtain

$$p_0'(t) = \mu e^{-(\lambda+\mu)t} + \lambda e^{-(\lambda+\mu)t} p_0^2(0) + \lambda \int_0^t e^{-(\lambda+\mu)s} \frac{d}{dt} p_0^2(t-s) ds \tag{4.42}$$

Equation (4.42) is then simplified as follows:

$$\begin{aligned}
p_0'(t) &= \mu e^{-(\lambda+\mu)t} + \lambda e^{-(\lambda+\mu)t} p_0^2(0) + \\
&\quad -\lambda \int_0^t e^{-(\lambda+\mu)s} \frac{d}{ds} p_0^2(t-s) ds = \\
&= \mu e^{-(\lambda+\mu)t} - \lambda e^{-(\lambda+\mu)s} p_0^2(t-s)\Big|_{s=0}^{s=t} + \\
&\quad -\lambda(\lambda+\mu) \int_0^t e^{-(\lambda+\mu)s} p_0^2(t-s) ds \\
&= \mu e^{-(\lambda+\mu)t} + \lambda p_0^2(t) + \\
&\quad -(\lambda+\mu)\left[p_0(t) - \mu \int_0^t e^{-(\lambda+\mu)s} ds\right] = \\
&= -(\lambda+\mu)p_0(t) + \lambda p_0^2(t) + \mu e^{-(\lambda+\mu)t} + \left[-\mu e^{-(\lambda+\mu)s}\right]_{s=0}^{s=t} \\
&= \mu - (\lambda+\mu)p_0(t) + \lambda p_0^2(t).
\end{aligned} \tag{4.43}$$

Hence the extinction probability $p_0(t)$, $t > 0$, satisfies the non-homogeneous Riccati equation

$$\frac{dp_0(t)}{dt} + (\lambda+\mu)p_0(t) = \lambda p_0^2(t) + \mu \tag{4.44}$$

An alternative derivation is the following.

By considering the process in the initial interval $[0, ds)$ and writing

$$p_0(t) = \mu ds + (1-\mu ds)(1-\lambda ds)p_0(t-ds) + \\ +(1-\mu ds)\lambda ds p_0^2(t-ds), \tag{4.45}$$

we are supposing that

1. the process becomes extincted in the initial interval $[0, ds)$ with probability μds;
2. the process survives and does not generate an additional offspring in the initial interval ds, and gets extincted in the interval $[ds, t)$ with probability
$$(1-\mu ds)(1-\lambda ds)p_0(t-ds); \tag{4.46}$$
3. the process generates an additional component with probability λds and in the remaining time $[ds, t)$, the two components (independently) disappear.

Neglecting infinitesimal of higher order we have that

$$\frac{p_0(t)-p_0(t-ds)}{ds} = \mu - (\lambda+\mu)p_0(t-ds) + \lambda p_0^2(t-ds) \tag{4.47}$$

and for $ds \to 0$ the Riccati equation emerges and is satisfied by the extinction probability (4.16) above.

Functions (4.21) and (4.22) are solutions of the Riccati equation of the next section.

For $\lambda = \mu$ we have that

$$\frac{d}{dt}p_0(t) + 2\mu p_0(t) = \mu p_0^2(t) + \mu \tag{4.48}$$

is satisfied by (4.22).

Finally we observe that

$$\lim_{t\to\infty} p_0(t) = \begin{cases} \frac{\mu}{\lambda}, & \lambda > \mu \\ 1, & \lambda = \mu \\ 1, & \lambda < \mu \end{cases} \tag{4.49}$$

according to intuition.

Note that the extinction is certain if $\lambda \leq \mu$ and it is still possible if $\lambda > \mu$.

Remark 4.4.
In the *general* birth-death process (see Remark 4.2) be $\rho(t) = \int_0^t [\mu(u) - \lambda(u)]du$, then its expected value is

$$\mathbb{E}[N(t)] = e^{-\rho(t)} \tag{4.50}$$

Its probability of extinction (Kendall, 1948; see also Tavaré, 2018) reads

$$\frac{\int_0^\infty e^{\rho(u)}\mu(u)du}{1 + \int_0^\infty e^{\rho(u)}\mu(u)du} \tag{4.51}$$

4.4 Galton-Watson Branching Processes

Introduction. We consider a population evolving in generations. We denote by Z_n the size of the n-th generation. Each member of the n-th generation generates a family of random size. The result of births of the n-th generation is the $(n+1)$-th generation.

The size of the successive generations is the sequence

$$Z_0, Z_1, ..., Z_n, ... \tag{4.52}$$

By $G_n(u)$ we denote the probability generating function of the r.v. Z_n, that is

$$\begin{aligned} G_n(u) = \mathbb{E}\left[u^{Z_n}\right] = \mathbb{E}\left[u^{X_1+...+X_{Z_{n-1}}}\right] = \\ = \mathbb{E}\left[\left(\mathbb{E}\left[u^X\right]\right)^{Z_{n-1}}\right] = \mathbb{E}[G(u)^{Z_{n-1}}] = \\ \text{(by posing } G(u) = v) \\ = \mathbb{E}\left[v^{Z_{n-1}}\right] = \mathbb{E}\left[G(v)^{Z_{n-2}}\right] = \\ = \mathbb{E}\left[G(G(u))^{Z_{n-2}}\right] = ... = \underbrace{G(G(G(...G(u))))}_{n} \end{aligned} \tag{4.53}$$

where G is the p.g.f. of the r.v. X ("offspring distribution").

The mean of the Galton-Watson branching process. Since the *probability generating function* has the nested structure

$$G_n(u) = G(G_{n-1}(u)) \tag{4.54}$$

its first derivative is

$$G'_n(u) = G'(G_{n-1}(u)) \cdot G'_{n-1}(u) \tag{4.55}$$

where, for $u = 1$, $G_{n-1}(u) = 1$, $G'(1) = \mathbb{E}[X]$ and hence

$$\mathbb{E}[Z_n] = \mathbb{E}[X] \cdot \mathbb{E}[Z_{n-1}] = \mathbb{E}[X](\mathbb{E}[X] \cdot \mathbb{E}[Z_{n-2}]) = ... = (\mathbb{E}[X])^n \tag{4.56}$$

The variance of the Galton-Watson branching process. Given the nested structure of the p.g.f. $G_n(u) = G(G_{n-1}(u))$ and its first derivative $G'_n(u) = G'(G_{n-1}(u)) \cdot G'_{n-1}(u)$, then its second derivative reads

$$G''_n(u) = G''(G_{n-1}(u))[G'_{n-1}(u)]^2 + G'(G_{n-1}(u))G''_{n-1}(u) \tag{4.57}$$

and

$$\begin{cases} G''_n(u)_{|u=1} = G''(1)[G'_{n-1}(1)]^2 + G'(1)G''_{n-1}(1) \\ G'_n(u)_{|u=1} = G'(1)G'_{n-1}(1) = G'(1)^n \end{cases} \tag{4.58}$$

Thus, from (4.58) we have that

$$\begin{aligned} \mathbb{E}[Z_n(Z_n - 1)] = \\ = \mathbb{E}[X(X-1)] \cdot (\mathbb{E}[X])^{2n-2} + \mathbb{E}[X] \cdot \mathbb{E}[Z_{n-1}(Z_{n-1} - 1)] \end{aligned} \tag{4.59}$$

that is

$$\begin{aligned} \mathbb{E}[Z_n^2] - \mathbb{E}[Z_n] = \\ = (\mathbb{E}[X^2] - \mathbb{E}[X]) \cdot (\mathbb{E}[X])^{2n-2} + \mathbb{E}[X] \cdot \left(\mathbb{E}[Z_{n-1}^2] - \mathbb{E}[Z_{n-1}]\right) \end{aligned} \tag{4.60}$$

which, from (4.56), becomes

$$\begin{aligned} \mathbb{E}[Z_n^2] - (\mathbb{E}[X])^n = \\ = \mathbb{E}[X^2] \cdot (\mathbb{E}[X])^{2n-2} - (\mathbb{E}[X])^{2n-1} + \mathbb{E}[X] \cdot \mathbb{E}[Z_{n-1}^2] - (\mathbb{E}[X])^n \end{aligned} \tag{4.61}$$

Finally, the variance obtains as

$$\begin{aligned} \mathbb{E}[Z_n^2] - (\mathbb{E}\,[Z_n])^2 = \\ = \left\{\mathbb{E}[X^2] - (\mathbb{E}[X])^2 + (\mathbb{E}[X])^2\right\} \cdot (\mathbb{E}[X])^{2n-2} + \\ -(\mathbb{E}[X])^{2n-1} - (\mathbb{E}[Z_n])^2 + \\ +\mathbb{E}[X] \cdot \left\{\mathbb{E}[Z_{n-1}^2] - (\mathbb{E}[Z_{n-1}])^2 + (\mathbb{E}[Z_{n-1}])^2\right\} \end{aligned} \tag{4.62}$$

that is, as the following recurrent relation, which is key to obtaining the final variance formula:

$$\begin{aligned} \mathbb{V}ar[Z_n] = \mathbb{V}ar[X] \cdot (\mathbb{E}[X])^{2n-2} + (\mathbb{E}[X])^{2n} - (\mathbb{E}[X])^{2n-1} + \\ -(\mathbb{E}[X])^{2n} + \mathbb{E}[X] \cdot \mathbb{V}ar[Z_{n-1}] + (\mathbb{E}[X])^{2n-1} = \\ = \mathbb{V}ar[X] \cdot (\mathbb{E}[X])^{2n-2} + \mathbb{E}[X] \cdot \mathbb{V}ar[Z_{n-1}] \end{aligned} \tag{4.63}$$

By successive substitutions, the final expression for the variance reads

$$\begin{aligned} \mathbb{V}ar[Z_n] = \sigma^2 \cdot \left\{\mu^{2n-2} + \ldots + \mu^{n-1}\right\} + \mu^n \mathbb{V}ar[Z_0] = \\ = \sigma^2 \cdot \mu^{n-1}\left\{1 + \ldots + \mu^{n-1}\right\} = \sigma^2 \cdot \mu^{n-1}\left(\frac{\mu^n - 1}{\mu - 1}\right) \end{aligned} \tag{4.64}$$

where μ and σ^2 are the mean and variance of the r.v. X (offspring distribution) and $\mathbb{V}ar[Z_0] = 0$.

4.5 Estimation of the Birth-death Process

This section provides an example of the estimation of a birth-and-death process on the population of bears in the Italian region 'Trentino'.

Continuous-time sampling scheme. Theorem 2.1 in Keiding (1975) provides the likelihood function of the simple (linear) birth-death process under continuous-time sampling, i.e.

$$L(\lambda, \mu|\boldsymbol{x}) \approx \lambda^{B_t}\mu^{D_t}\exp\left(-(\lambda+\mu)S_t\right) \tag{4.65}$$

with $S_t = \int_0^t X_u du$, B_t = number of births, D_t = number of deaths, and it provides the maximum likelihood estimators, i.e.

$$\hat{\lambda} = \frac{B_t}{S_t} \qquad \hat{\mu} = \frac{D_t}{S_t} \tag{4.66}$$

Remark 4.5.
Let $N_t = B_t + D_t$ denote the total number of events; then the maximum likelihood estimators of the quantities $\lambda+\mu$ and $\lambda-\mu$ (the Malthusian growth rate) are, respectively,

$$\widehat{\lambda+\mu} = \frac{N_t}{S_t} \qquad \widehat{\lambda-\mu} = \frac{B_t - D_t}{S_t} = \frac{X_t - x_0}{S_t} \tag{4.67}$$

Note that B_t events correspond each to a +1 (up) jump and D_t events to a −1 (down) jump, and the net displacement is $X_t - x_0 = B_t - D_t$.

Theorem 4.2. *As $x_0 \to \infty$, it holds that*

$$(\hat{\lambda}, \lambda) \to (\lambda, \mu) \qquad a.s. \tag{4.68}$$

and

$$\sqrt{x_0 \frac{(e^{(\lambda-\mu)t} - 1)}{\lambda - \mu}} \begin{pmatrix} \hat{\lambda} - \lambda \\ \hat{\mu} - \mu \end{pmatrix} \to \mathcal{N}\left(\begin{pmatrix} 0 \\ 0 \end{pmatrix}, \begin{pmatrix} \lambda & 0 \\ 0 & \mu \end{pmatrix} \right) \tag{4.69}$$

If $\lambda = \mu$ the factor is simply $\sqrt{x_0 t}$.

Proof. See Keiding, 1975, Theor. 3.1. □

Theorem 4.3. *As $t \to \infty$, it holds that*

$$\frac{e^{-(\lambda-\mu)t}}{x_0} \begin{pmatrix} B_t \\ D_t \\ S_t \end{pmatrix} \to \begin{pmatrix} \lambda \\ \mu \\ 1 \end{pmatrix} \frac{W}{\lambda - \mu} \tag{4.70}$$

where W is $\Gamma(x_0, \frac{1}{x_0})$, with density $f_\Gamma(w) = \frac{x_0^{x_0} w^{x_0 - 1} e^{-x_0 w}}{\Gamma(x_0)}$.

Proof. See Keiding, 1975, Theor. 4.1. □

Discrete sampling scheme. Assume n observations (plus 1 progenitor) $\boldsymbol{x} \equiv (x_0, x_1, ..., x_n)$ are sampled over a time period $[0, T)$ at equidistant time points t_i, $i = 0, 1, 2, ..., n$ at a time step $\Delta t = \frac{T}{n}$, such that $t_0 = 0$, $t_n = T$ and $X(t_i) = x_i$.

The maximum likelihood estimator of the *Malthusian* growth rate $\lambda - \mu$ is given by (see Theor. 6.1 in Keiding, 1975)

$$\widehat{\lambda - \mu} = \frac{1}{\Delta t} \log \left(\frac{x_1 + x_2 + ... + x_n}{x_0 + x_1 + ... + x_{n-1}} \right) \tag{4.71}$$

Remark 4.6.
For the Malthusian rate $\lambda - \mu$, given a sample of n observations equally spaced at times $t_i = t_{i-1} + \Delta t$ ($\Delta t > 0$), a moment-method estimator is available in closed form from the relation $\mathbb{E}[N(t) \mid N(t_{i-1})] = N(t_{i-1})e^{+(\lambda-\mu)t}$

$$\widehat{\lambda - \mu} = \frac{1}{\Delta t} \log \left(\frac{\sum_{h=1}^{n} x_h x_{h-1}}{\sum_{h=0}^{n-1} x_h^2} \right) \tag{4.72}$$

where x_i denotes the sampled value of $N(t_i)$ (see Tavaré, 2018).

Davison et Al. (2021) propose an estimator based on an "embedded" Galton-Watson process.

A Galton-Watson process is a *branching process* X_n that tracks the number of offsprings or descendants from one progenitor ($X(0) = 1$) and it was introduced by Francis Galton in the statistical study of the extinction of family names.

Its mean and variance are m and σ^2, and their relation with the birth and death rates is (see Davison et Al., 2021)

$$\begin{cases} \lambda = \frac{\log(m)}{2\Delta t} \left(\frac{\sigma^2}{m(m-1)} + 1 \right) \\ \mu = \frac{\log(m)}{2\Delta t} \left(\frac{\sigma^2}{m(m-1)} - 1 \right) \end{cases} \tag{4.73}$$

Hence, given a sample of n observations $\underline{x} \equiv \{x_1, ..., x_n\}$, the estimators proposed in Davison et Al. (2021) read:

$$\begin{cases} \hat{m} = \frac{\sum_{i=1}^{n} x_i}{\sum_{i=1}^{n} x_{i-1}} \\ \hat{\sigma^2} = \frac{1}{n}\sum_{i=1}^{n} x_{i-1}\left(\frac{x_i}{x_{i-1}} - \hat{m}\right)^2 \end{cases} \tag{4.74}$$

Note that Davison et al. (2021) refer to $\hat{m}$ as a method-of-moments, conditional least square and non parametric maximum likelihood estimator of the offspring mean. Note further that it implicitly also depends (through the element x_0 in the denominator of the first summand) on the initial number n_0 of progenitors.

The resulting estimators for the birth and death rates are:

$$\begin{cases} \hat{\lambda} = \frac{\log\{\hat{m}\}}{2\Delta t}\left(\frac{\hat{\sigma^2}}{\hat{m}(\hat{m}-1)} + 1\right) \\ \hat{\mu} = \frac{\log\{\hat{m}\}}{2\Delta t}\left(\frac{\hat{\sigma^2}}{\hat{m}(\hat{m}-1)} - 1\right) \end{cases} \tag{4.75}$$

As for the asymptotic properties of the estimators, asymptotically for $n \to \infty$ the following limit holds (see theorem 1 in Davison, 2021):

$$\sqrt{n}\begin{pmatrix} \hat{\lambda}_n - \lambda \\ \hat{\mu}_n - \mu \end{pmatrix} \sim \mathcal{N}\left(\begin{pmatrix} 0 \\ 0 \end{pmatrix}, \frac{(\log(m))^2\sigma^4}{2\Delta t^2 m^2(m-1)^2} \times \begin{pmatrix} 1, 1 \\ 1, 1 \end{pmatrix}\right) \tag{4.76}$$

where n (here) is the number of years; be $\omega = \lambda - \mu$, again, for $n \to \infty$ the following limit holds (see theorem 2 in Davison, 2021):

$$\sqrt{\sum_{j=1}^{n} X_j} \cdot (\hat{\omega}_n - \omega) \sim \mathcal{N}\left(0, \frac{\sigma^2}{m^2(\Delta t)^2}\right) \tag{4.77}$$

Example. The population of bears in the Trentino region (Italy)

The example considers the population of horses in the region of Trentino (Italy). In Italy, as of year 2021, there are about 200 horses, approximately concentrated in the Alps (Trentino and

TABLE 4.1
Evolution of the bears population of Trentino

Bears in Trentino: Population size, number of births and deaths			
year	population	births	deaths
2000	1	.	.
2001	9	8	0
2002	11	2	0
2003	12	2	1
2004	16	5	1
2005	19	4	1
2006	26	11	4
2007	24	3	5
2008	28	8	4
2009	30	4	2
2010	34	9	5
2011	37	8	5
2012	49	20	8
2013	48	6	7
2014	51	13	10
2015	54	13	10
2016	63	11	2
2017	70	11	4
2018	78	21	13
2019	95	17	0
2020	114	22	3
2021	121	12	5
2022	143	25	3

Friuli) and Appennini (Abruzzo) mountains. A third of them are in Trentino, as a result of a campaign to prevent its extinction. A monitoring is established since 2002, such that the population size, births[2] and deaths are yearly recorded[3].

[2]Births include repopulation. The initial population of bears has been approximated to $X(0) = 1$ (see Davison et al., 2021).
[3]Source: Trento Province https://grandicarnivori.provincia.tn.it/

Hence, $\hat{m} = 1.143$ and $\hat{\sigma^2} = 3.173$ and

$$\hat{\lambda} = \frac{\log\{1.\hat{1}43\}}{2\Delta t}\left(\frac{3.\hat{1}73}{1.\hat{1}43(1.\hat{1}43-1)} + 1\right) = 0.592$$

$$\hat{\mu} = \frac{\log\{1.\hat{1}43\}}{2\Delta t}\left(\frac{3.\hat{1}73}{1.\hat{1}43(1.\hat{1}43-1)} - 1\right) = 0.534$$

Part III

Fractional Poisson Processes

5

Fractional Extensions of the Poisson and Birth and Death Processes

CONTENTS

This chapter is devoted to fractional extensions of the Poisson process, with emphasis on two members of this class, namely the space-fractional and time-fractional Poisson processes, with a discussion of their probabilistic properties. First the space-fractional Poisson process, introduced by Orsingher and Polito (2012b), is presented, for which the state distribution is derived, and explicit state probabilities are reported for small values of k. Then, two probabilistic models built on the time-fractional Poisson process are presented, together with their probabilistic properties. For all processes, their link to subordination and time-changes are discussed. Finally, the time-fractional Poisson process is implemented using earthquakes data, estimating its fractional parameter and ruin probability, based on results in Constantinescu et al. (2019). Moreover, the chapter introduces the main results obtained

DOI: 10.1201/9781003398011-5

by Orsingher and coauthors (2010, 2011) on the fractional counterpart of the pure-birth, pure-death and birth-death processes, where fractionality is obtained by replacing the time derivative in the difference-differential equation of the probability law with the Dzherbashyan-Caputo fractional derivative.

5.1 The Space-fractional Poisson Process: Standard Case

It is well-known that the state probabilities of the homogeneous Poisson process satisfy the difference-differential equations

$$\frac{dp_k(t)}{dt} = -\lambda p_k(t) + \lambda p_{k-1}(t) = -\lambda(I-B)p_k(t) \qquad k \geq 0 \quad (5.1)$$

where $Ip_k = p_k$ and $Bp_k = p_{k-1}$ where B is the "shift" operator.

The space-fractional Poisson process is defined by considering that the state probabilities (see Orsingher and Polito, 2012b) satisfy

$$\frac{dp_k^{(\alpha)}(t)}{dt} = -[\lambda(I-B)]^{\alpha} p_k^{(\alpha)}(t), \quad 0 < \alpha < 1 \quad k \geq 0 \qquad (5.2)$$

so that we arrive at the so-called space-fractional Poisson process, denoted as $N^{\alpha}(t), t > 0$.

Equation (5.2) can be rewritten by expanding the operator $(I-B)^{\alpha}$ in the binomial series, because the binomial expansion reads

$$(1-x)^{\alpha} = \sum_{h=0}^{\infty} \frac{\Gamma(\alpha+1)}{h!\Gamma(\alpha+1-h)}(-x)^h 1^{\alpha-h}, \quad |x| < 1 \qquad (5.3)$$

where $\frac{\Gamma(\alpha+1)}{\Gamma(\alpha+1-h)}$ corresponds to the s.c. Pochhammer symbol $(\alpha)_h$ so that

$$(I-B)^{\alpha} = \sum_{h=0}^{\infty} \frac{\Gamma(\alpha+1)}{h!\Gamma(\alpha+1-h)}(-B)^h I^{\alpha-h} \qquad (5.4)$$

and, substituting in equation (5.2)

$$\frac{dp_k^{(\alpha)}(t)}{dt} = -\lambda^\alpha \sum_{h=0}^{k} \frac{\Gamma(\alpha+1)}{h!\Gamma(\alpha+1-h)}(-1)^h p_{k-h}^{(\alpha)}(t) \tag{5.5}$$

subject to the initial conditions, that is

$$p_k^{(\alpha)}(0) = \begin{cases} 1 & k=0 \\ 0 & k \geq 1 \end{cases} \tag{5.6}$$

5.1.1 Properties of the Space-fractional Poisson Process

(i) Note that $\frac{dp_k^{(\alpha)}(t)}{dt}$ in (5.5) depends on all probabilities $p_{k-h}(t)$ with $0 \leq h \leq k$. This means that the jumps of the space-fractional Poisson process can be of any integer-valued size (see Fig. 5.1).

(ii) For $\alpha = 1$ equation (5.5) reduces to equation (5.1).

(iii) **The local distribution of jumps**. The distribution of jumps (over an infinitesimal time interval dt) can be written as

$$\Pr\{N^\alpha(t, t+dt] = k\} = \\ = \begin{cases} \frac{(-1)^{k+1}\lambda^\alpha\Gamma(\alpha+1)}{k!\Gamma(\alpha+1-k)}dt + o(dt) & k \geq 1 \\ 1 - \lambda^\alpha dt + o(dt) & k = 0 \end{cases} \tag{5.7}$$

where $(\alpha)_k = \alpha(\alpha-1)\ldots(\alpha+1-k)$ denotes the Pochhammer symbol.

Alternatively the distribution (5.7) can be rewritten as

$$\Pr\{N^\alpha(t, t+dt] = k\} = \begin{cases} dt\frac{\lambda^k}{k!}\int_0^\infty e^{-\lambda s}s^k\nu(ds) + o(dt), & k \geq 1 \\ 1 - dt\int_0^\infty(1-e^{-\lambda s})\nu(ds) + o(dt), & \\ & k = 0 \end{cases} \tag{5.8}$$

where $\nu(ds) = \frac{\alpha s^{-\alpha-1}}{\Gamma(1-\alpha)}ds, s > 0$ is the Lévy measure that is a measure on the half-line such that

$$\int_0^\infty (s \wedge 1)\,\nu(ds) < \infty \tag{5.9}$$

Indeed, for the Lévy measure $\nu(ds) = \frac{\alpha s^{-\alpha-1}}{\Gamma(1-\alpha)}$, we show that $\frac{\lambda^k}{k!}\int_0^\infty e^{-\lambda s}s^k\nu(ds) = \frac{(-1)^{k+1}\lambda^\alpha\Gamma(\alpha+1)}{k!\Gamma(\alpha+1-k)}$.

In fact,

$$
\begin{aligned}
\frac{\lambda^k}{k!}\int_0^\infty e^{-\lambda s}s^k\nu(ds) = & \\
&= \frac{\lambda^k}{k!}\int_0^\infty e^{-\lambda s}s^k\frac{\alpha s^{-\alpha-1}}{\Gamma(1-\alpha)}ds = \\
&= \frac{\lambda^k\alpha}{k!\Gamma(1-\alpha)}\int_0^\infty e^{-\lambda s}s^{k-\alpha-1}ds = \\
&= \frac{\lambda^k\alpha}{k!\Gamma(1-\alpha)}\frac{\Gamma(k-\alpha)}{\lambda^{k-\alpha}} = \\
&= \frac{\lambda^k\alpha}{k!}\frac{\Gamma(\alpha)\sin(\pi\alpha)}{\lambda^{k-\alpha}}\times \\
&\times\frac{\pi}{\pi\Gamma(1-(k-\alpha))\sin(\pi(k-\alpha))} = \\
&= \frac{\alpha\Gamma(\alpha)\lambda^\alpha}{k!}\frac{\sin(\pi\alpha)}{\Gamma(1-(k-\alpha))\sin(\pi(k-\alpha))} = \\
&= (-1)^{k+1}\frac{\Gamma(\alpha+1)\lambda^\alpha}{k!\Gamma(1-(k-\alpha))}
\end{aligned}
\tag{5.10}
$$

where Euler's *reflection formula* (see Appendix D.2) of the gamma function has been applied twice, and the addition formula for the sine function is used as well.

(iv) The probability generating function of the space-fractional Poisson process $N^\alpha(t), t > 0$, as a consequence of the independence of increments, reproduces the multiplicative structure of the standard Poisson process.

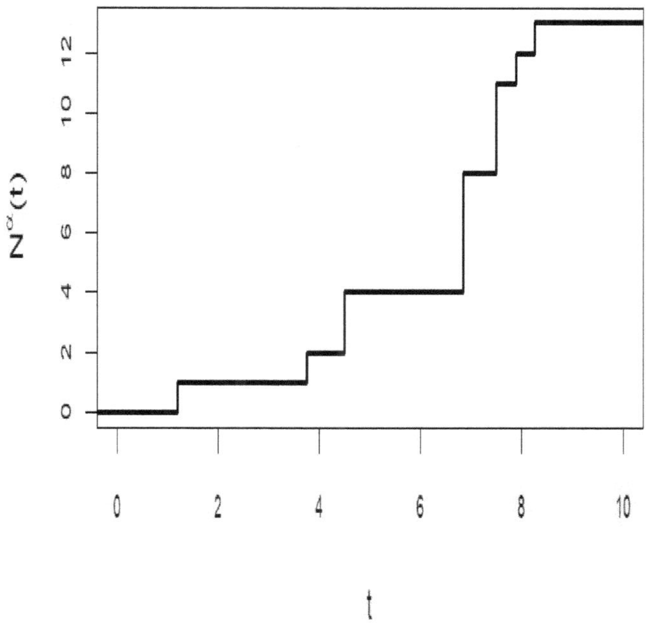

FIGURE 5.1
Sample path for $\alpha < 1$ of a space-fractional Poisson process $N^\alpha(t)$.

Theorem 5.1. *The probability generating function of* $N^\alpha(t)$ *satisfies the equation*

$$\begin{cases} \frac{d}{dt}G_\alpha(u,t) &= -\lambda^\alpha(1-u)^\alpha G_\alpha(u,t) \\ G_\alpha(u,0) &= 1 \end{cases} \tag{5.11}$$

so the p.g.f. is obtained by solving the (5.11)*:*

$$G_\alpha(u,t) = e^{-[\lambda(1-u)]^\alpha t} \tag{5.12}$$

Proof. From (5.5) we have that

$$\sum_{k=0}^{\infty} u^k \frac{dp_k^{(\alpha)}(t)}{dt} = -\lambda^\alpha \sum_{k=0}^{\infty} u^k \sum_{h=0}^{k} \frac{\Gamma(\alpha+1)}{h!\Gamma(\alpha+1-h)}(-1)^h p_{k-h}^{(\alpha)}(t) \tag{5.13}$$

i.e.

$$\begin{aligned}
\frac{\partial}{\partial t}G_\alpha(u,t) &= \\
&= -\lambda^\alpha\Gamma(\alpha+1)\sum_{h=0}^{\infty}\sum_{k=h}^{\infty}\frac{(-1)^h}{h!\Gamma(\alpha+1-h)}p_{k-h}^\alpha(t)\cdot u^k = \\
&= -\lambda^\alpha\Gamma(\alpha+1)\sum_{h=0}^{\infty}\sum_{k=0}^{\infty}\frac{(-1)^h u^{k+h}}{h!\Gamma(\alpha+1-h)}p_k^\alpha(t) = \\
&= -\lambda^\alpha\Gamma(\alpha+1)G_\alpha(u,t)\cdot\sum_{h=0}^{\infty}\frac{(-1)^h}{h!\Gamma(\alpha+1-h)}u^h = \\
&= -\lambda^\alpha G_\alpha(u,t)(1-u)^\alpha
\end{aligned} \tag{5.14}$$

because

$$\begin{aligned}
(1-u)^\alpha &= \sum_{h=0}^{\infty}\binom{\alpha}{h}(-u)^h = \\
= \sum_{h=0}^{\infty}\frac{\Gamma(\alpha+1)}{h!\Gamma(\alpha-h+1)}(-1)^h u^h &= \sum_{h=0}^{\infty}\frac{\Gamma(\alpha+1)(-u)^h}{h!\Gamma(\alpha-h+1)}
\end{aligned} \tag{5.15}$$

Thus the solution to equation (5.11), with the given initial condition, is

$$G(u,t) = Ke^{-\lambda^\alpha(1-u)^\alpha t} \tag{5.16}$$

and $K = 1$ since $G(u,0) = 1$. □

Moreover, the following property also holds true.

Theorem 5.2. *The p.g.f. (5.12) is itself a probability and can be written as:*

$$G_\alpha(u,t) = e^{-[\lambda(1-u)]^\alpha t} = Pr\left\{\min_{1\le k\le N(t)} X_k^{\frac{1}{\alpha}} \ge 1-u\right\} \quad 0<u<1 \tag{5.17}$$

where the X_k's are independent r.v.'s with Uniform distribution in $[0,1]$, $N(t)$ is a homogeneous Poisson process of rate λ^α.

Proof. Since

$$\Pr\left\{\min_{1\le k\le N(t)} X_k^{\frac{1}{\alpha}} \ge 1-u\right\} = \\ = \mathbb{E}\left[\Pr\left\{\min_{1\le k\le N(t)} X_k^{\frac{1}{\alpha}} \ge 1-u\right\} \,\middle|\, N(t)\right] \tag{5.18}$$

and by observing that

$$\Pr\left(\min_{0\le k\le N(t)} X_k^{\frac{1}{\alpha}} \ge 1-u \,\middle|\, N(t)=n\right) = \\ = \Pr\left(\min_{0\le k\le n} X_k^{\frac{1}{\alpha}} \ge 1-u\right) = \Pr\left(\prod_{k=1}^{n}[X_k \ge (1-u)^\alpha]\right) = \\ = \prod_{k=1}^{n}\Pr\{X_k \ge (1-u)^\alpha\} = [\Pr\{X_k \ge (1-u)^\alpha\}]^n = \\ = \left[\int_{(1-u)^\alpha}^{1} dx\right]^n = [1-(1-u)^\alpha]^n \tag{5.19}$$

the result follows, i.e.

$$\Pr\left(\min_{0\le k\le N(t)} X_k^{\frac{1}{\alpha}} \ge 1-u\right) = \\ = \mathbb{E}\left[\Pr\{\min_{0\le k\le N(t)} X_k^{\frac{1}{\alpha}} \ge 1-u\} \,\middle|\, N(t)\right] = \\ = \sum_{n=0}^{\infty}\Pr\left(\min_{0\le k\le N(t)} X_k^{\frac{1}{\alpha}} \ge 1-u \,\middle|\, N(t)=n\right)\cdot \Pr\{N(t)=n\} = \\ = \sum_{n=0}^{\infty}[1-(1-u)^\alpha]^n \cdot e^{-\lambda^\alpha t}\frac{(\lambda^\alpha t)^n}{n!} = \\ = e^{-\lambda^\alpha t}\sum_{n=0}^{\infty}\frac{[1-(1-u)^\alpha]^n\,(\lambda^\alpha t)^n}{n!} = \\ = e^{-\lambda^\alpha t}\cdot \exp\{[1-(1-u)^\alpha]\lambda^\alpha t\} = \\ = e^{-\lambda^\alpha (1-u)^\alpha t} \tag{5.20}$$

□

(v) By virtue of the properties of the exponential function in (5.12) the space-fractional process has independent increments.

In fact

$$
\begin{aligned}
G_\alpha(u,t) = e^{-\lambda^\alpha(1-u)^\alpha t} &= e^{-\lambda^\alpha(1-u)^\alpha[s+t-s]} = \\
&= G_\alpha(u,s)\cdot G_\alpha(u,t-s)
\end{aligned}
\tag{5.21}
$$

(vi) From (5.12) the probability law of $N^\alpha(t)$ can be obtained.

Theorem 5.3. *For $0 \le \alpha \le 1$ the distribution of the space-fractional Poisson process $N^\alpha(t)$ is*

$$
p_k^\alpha(t) = Pr\{N^\alpha(t)=k\} = \frac{(-1)^k}{k!}\sum_{h=0}^{\infty}\frac{(-\lambda^\alpha t)^h}{h!}\frac{\Gamma(\alpha h+1)}{\Gamma(\alpha h+1-k)}, \quad k \ge 0.
\tag{5.22}
$$

For $\alpha = 1$ immediately emerges that (5.22) coincides with the classical Poisson distribution.

Proof. By expanding the probability-generating function $G_\alpha(u,t)$ (5.12), the state probabilities obtain as the coefficients of the expansion, i.e.

$$
\begin{aligned}
G_\alpha(u,t) &= e^{-\lambda^\alpha(1-u)^\alpha t} = \\
= \sum_{h=0}^{\infty}\frac{[-\lambda^\alpha(1-u)^\alpha t]^h}{h!} &= \sum_{h=0}^{\infty}\frac{(-\lambda^\alpha t)^h(1-u)^{\alpha h}}{h!} = \\
&= \sum_{h=0}^{\infty}\frac{(-\lambda^\alpha t)^h}{h!}\cdot\sum_{k=0}^{\infty}\frac{\Gamma(\alpha h+1)}{k!\Gamma(\alpha h+1-k)}(-u)^k = \\
&= \sum_{k=0}^{\infty}\frac{(-1)^k}{k!}\sum_{h=0}^{\infty}\frac{(-\lambda^\alpha t)^h}{h!}\frac{\Gamma(\alpha h+1)}{\Gamma(\alpha h+1-k)}\cdot u^k
\end{aligned}
\tag{5.23}
$$

and, hence,

$$
p_k^\alpha(t) = \frac{(-1)^k}{k!}\sum_{h=0}^{\infty}\frac{(-\lambda^\alpha t)^h}{h!}\frac{\Gamma(\alpha h+1)}{\Gamma(\alpha h+1-k)}
\tag{5.24}
$$

For $\alpha = 1$ then $p_k^\alpha(t)$ becomes

$$
\begin{aligned}
p_k^1(t) = \frac{(-1)^k}{k!} \sum_{h=k}^{\infty} \frac{(-\lambda t)^k}{k!} \frac{h!}{(h-k)!} &= \\
= \frac{(-1)^k}{k!} \sum_{h'=0}^{\infty} \frac{(-\lambda t)^{h'+k}}{h'!} &= \\
= \frac{(-1)^k (-\lambda t)^k}{k!} \sum_{h=0}^{\infty} \frac{(-\lambda t)^h}{h!} &= \\
= \frac{(\lambda t)^k}{k!} e^{-\lambda t} &
\end{aligned}
\tag{5.25}
$$

□

Note that:

$$
\begin{aligned}
p_0^\alpha(t) &= e^{-\lambda^\alpha t} \\
p_1^\alpha(t) &= \alpha \lambda^\alpha t e^{-\lambda^\alpha t} \\
p_2^\alpha(t) &= \frac{e^{-\lambda^\alpha t}}{2!} \left[(\alpha \lambda^\alpha t)^2 + \alpha(1-\alpha)\lambda^\alpha t \right] \\
p_3^\alpha(t) &= \frac{e^{-\lambda^\alpha t}}{3!} \left[(\alpha \lambda^\alpha t)^3 + 3(\alpha \lambda^\alpha t)^2 (1-\alpha) + \alpha(1-\alpha)(2-\alpha)\lambda^\alpha t \right] \\
p_4^\alpha(t) &= \frac{e^{-\lambda^\alpha t}}{4!} \left[(\alpha \lambda^\alpha t)^4 + 6(\alpha \lambda^\alpha t)^3 (1-\alpha) + \right. \\
&\quad \left. +6(\alpha \lambda^\alpha t)^2 (1-\alpha)(2-\alpha) + \alpha(1-\alpha)(2-\alpha)(3-\alpha)\lambda^\alpha t \right]
\end{aligned}
\tag{5.26}
$$

(see Fig. 5.2 and Fig. 5.3) and in general:

$$
p_k^\alpha(t) = \frac{e^{-\lambda^\alpha t}}{k!} \left[c_{k,k} t^k + c_{k-1,k} t^{k-1} + \ldots + c_{2,k} t^2 + c_{1,k} t \right] \tag{5.27}
$$

See Orsingher and Toaldo (2015) for expressions of the coefficients.

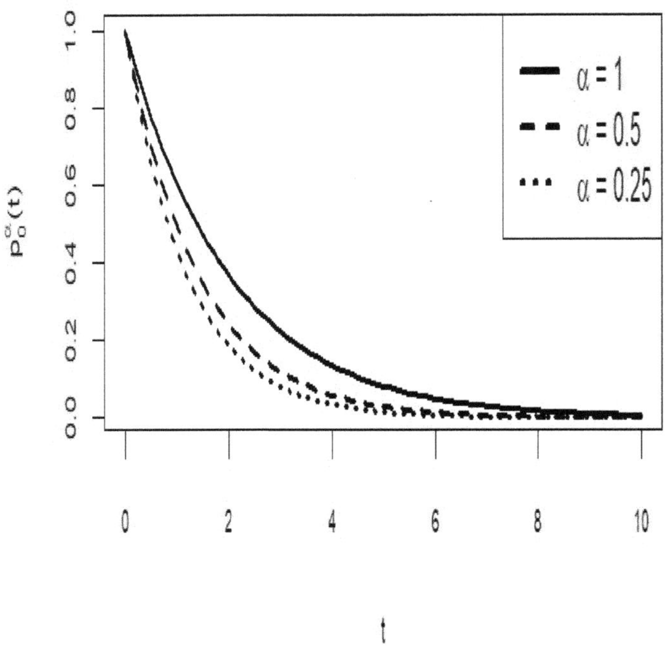

FIGURE 5.2
Space-fractional Poisson: The time evolution for $\alpha = 1$, $\alpha = 0.5$ and $\alpha = 0.25$ of the (survival) probability $p_0^\alpha(t)$, i.e. $k = 0$ ($\lambda = 0.5$).

(vii) Mean and variance diverge. In fact the first moment can be obtained from the first derivative of the p.g.f.

$$\begin{aligned} \frac{\partial}{\partial u}\mathbb{E}\left[u^{N^\alpha(t)}\right] = \mathbb{E}\left[N^\alpha(t)u^{N^\alpha(t)-1}\right]|_{u=1} = \mathbb{E}[N^\alpha(t)], \\ \frac{\partial}{\partial u}G_\alpha(u,t) = \frac{\partial}{\partial u}e^{-\lambda^\alpha(1-u)^\alpha t} = \\ = G_\alpha(u,t)(\alpha\lambda^\alpha t(1-u)^{\alpha-1})|_{u=1} = \infty \end{aligned} \tag{5.28}$$

being $\alpha < 1$.

The second moment can be obtained from the k-th factorial moment

$$\mu_{(k)}^X = \mathbb{E}[X(X-1)...(X-k+1)] \tag{5.29}$$

$$\mathbb{E}[X^2] = \mu_{(1)}^X + \mu_{(2)}^X \tag{5.30}$$

and the second moment diverges as well.

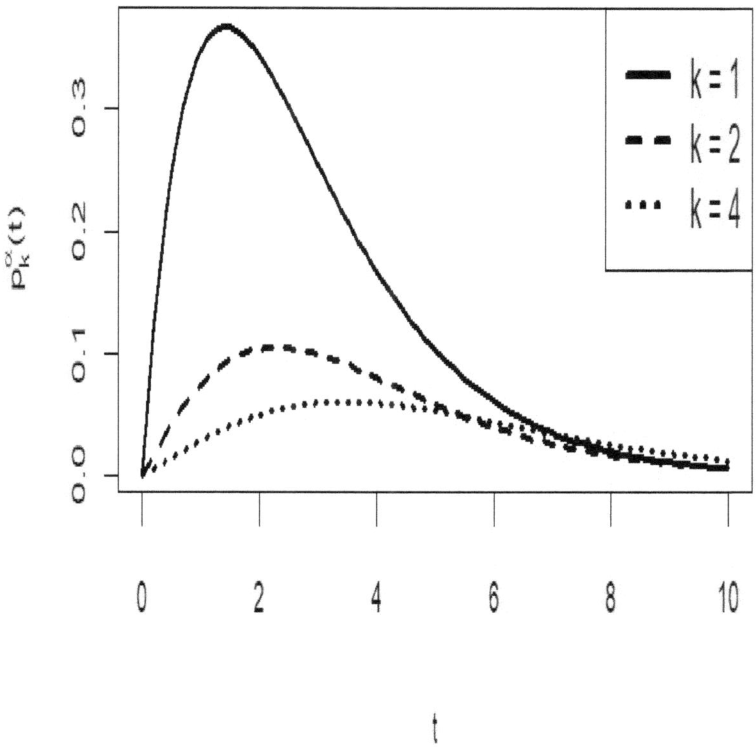

FIGURE 5.3
Space-fractional Poisson: The time evolution of the state probabilities $p_k^\alpha(t)$ for $k = 1$, $k = 2$ and $k = 4$, and $\alpha = 0.5$ $(\lambda = 0.5)$ (see Fig. 5.2).

(viii) The space-fractional Poisson process is a time-changed Poisson process $N^\alpha(t)$:

$$N^\alpha(t) \overset{D}{\sim} N(S_\alpha(t)) \tag{5.31}$$

where $S_\alpha(t)$ is a stable subordinator of order $0 < \alpha < 1$, independent from $N(t)$, that is a process with Laplace transform (see Appendix B)

$$\mathbb{E}\left[e^{-\gamma S_\alpha(t)}\right] = e^{-t\gamma^\alpha} \tag{5.32}$$

It is simple to check (5.31), having in mind (5.32), as follows

$$\begin{aligned} \mathbb{E}\left[u^{N(S^\alpha(t))}\right] &= \mathbb{E}\left[\mathbb{E}\left(u^{N(S^\alpha(t))}\middle|S^\alpha(t)\right)\right] = \mathbb{E}\left[e^{-\lambda(1-u)S^\alpha(t)}\right] = \\ &= e^{-t[\lambda(1-u)]^\alpha} = \mathbb{E}\left[u^{N^\alpha(t)}\right] = G_\alpha(u,t), \text{ for } |u| \le 1 \end{aligned} \tag{5.33}$$

(ix) Due to the possibility for the space-fractional Poisson process of multiple jumps over any interval dt, the probabilistic nature of waiting times is different from the Exponential distribution characterizing the homogeneous Poisson process. The first-passage time[1] of level k is defined as:

$$T_k^{(\alpha)} = \inf\{t \geq 0 : N^{(\alpha)}(t) \geq k\} \tag{5.34}$$

The distribution of $T_k^{(\alpha)}$ is obtained in Orsingher and Toaldo (2015) as

$$\Pr\{T_k^{(\alpha)} \in ds\} = -\frac{d}{ds}\sum_{h=0}^{k-1} \frac{(-\lambda)^h}{h!}\frac{d^h}{d\lambda^h}e^{-s\lambda^\alpha}, \quad s > 0 \tag{5.35}$$

where, for instance

$$\Pr\{T_k^{(\alpha)} \in ds\} = \begin{cases} \lambda^\alpha e^{-\lambda^\alpha s} ds & k = 1 \\ \lambda^\alpha e^{-\lambda^\alpha s}(1 - \alpha + \alpha\lambda^\alpha s) ds & k = 2 \end{cases} \tag{5.36}$$

Orsingher and Toaldo (2015) also derive an iterative construction, i.e.

$$\Pr\{T_k^{(\alpha)} \in ds\} = \Pr\{T_{k-1}^{(\alpha)} \in ds\} - \frac{(-\lambda)^{k-1}}{(k-1)!}\frac{d}{ds}\left(\frac{d^{k-1}}{d\lambda^{k-1}}e^{-\lambda^\alpha s}\right) ds. \tag{5.37}$$

[1]Note that, in the case of the space-fractional Poisson process, the first-passage time, as defined by equation (5.23), differs from the concept of *hitting time*, defined as $\hat{T}_k^{(\alpha)} = \inf\{t \geq 0 : N^{(\alpha)}(t) = k\}$, for which $\Pr\{\hat{T}_k^{(\alpha)} < \infty\} < 1$ and there is a positive probability of missing any level k (due to multiple jumps), see Garra, Orsingher and Scavino (2017).

Remark 5.1.
The space-fractional Skellam process is defined as

$$S(t) = N_1(S_{\alpha_1}(t)) - N_2(S_{\alpha_2}(t)) \tag{5.38}$$

with S_{α_1} and S_{α_2} subordinators independent of $N_1(t)$ and $N_2(t)$.
It can perform upward and downward jumps of arbitrary lengths. For its properties see Buchak, Sakhno (2019).

5.2 Generalized Space-fractional Poisson Processes

If we start from equation

$$\frac{dp_k(t)}{dt} = -f\left(\lambda\left(I - B\right)\right) p_k(t) \tag{5.39}$$

with $p_k(0) = 1$ for $k = 0$ and $p_k(0) = 0$ for $k \geq 1$, which generalizes (5.1) and (5.2), we are able to construct a new class of point processes with independent increments.

The functions f appearing in (5.39) must belong to the class of Bernštein functions, which are $C^\infty(\mathbb{R}^+)$, non-negative and such that[2]

$$(-1)^k \frac{d^k f(x)}{dx^k} \leq 0, \quad x > 0, k \geq 1 \tag{5.40}$$

Furthermore the Bernštein functions have the following integral representation (cf. Schilling et al., 2012)

$$f(x) = a + bx + \int_0^\infty (1 - e^{-xs})\nu(ds) \tag{5.41}$$

where $\nu(ds)$ is the Lévy measure on $(0, +\infty)$ and such that

$$\int_0^{+\infty} (s \wedge 1)\nu(ds) < \infty \tag{5.42}$$

[2]Note that the function $f(x)$ is such that $f'(x) \sim \int_0^\infty e^{-xs}\nu(ds) > 0$ and $f''(x) \sim -\int_0^\infty e^{-xs}\nu(ds) < 0$.

In our text we assume $a = b = 0$. For each Lévy measure we define a Bernštein function by means of (5.41).

Some particular Bernštein functions that are related to the processes of this paragraph are

$$\begin{aligned} f(x) &= x && (\textit{homogeneus Poisson}) \\ f(x) &= x^{\alpha} && (\textit{space-fractional Poisson}) \\ f(x) &= (x+\theta)^{\alpha} - \theta^{\alpha} && (\textit{tempered Poisson process}) \\ f(x) &= \log(1+x) && (\textit{negative binomial process}) \end{aligned} \tag{5.43}$$

The function x^{α} is related to by the Lévy measure

$$\nu(ds) = \frac{\alpha s^{-(\alpha+1)}}{\Gamma(1-\alpha)} ds, \quad 0 < \alpha < 1. \tag{5.44}$$

The function $(x+\theta)^{\alpha} - \theta^{\alpha}$ generates a process (tempered Poisson) with *finite* mean and variance (see Orsingher and Toaldo, 2015).

The process related to (5.39), with $t > 0$, is denoted by $N^f(t)$ and its probabilistic behaviour is governed by the following relationship

$$\begin{aligned} & Pr\left\{N^f(t, t+dt] = k\right\} = \\ &= \begin{cases} dt\frac{\lambda^k}{k!}\int_0^{\infty} e^{-\lambda s} s^k \nu(ds) + o(dt), & k \geq 1 \\ 1 - dt f(\lambda), & k = 0 \end{cases} \end{aligned} \tag{5.45}$$

The relationship with the homogeneous Poisson $N(t)$ is given by

$$N^f(t) = N(H^f(t)) \tag{5.46}$$

where $H^f(t)$ is a subordinator with Laplace transform

$$\mathbb{E}\left[e^{-\mu H^f(t)}\right] = e^{-tf(\mu)} = e^{-t\int_0^{\infty}(1-e^{-\mu s})\nu(ds)} \tag{5.47}$$

The probability generating function of N^f can be derived from (5.39), i.e.

$$\begin{cases} \frac{\partial}{\partial t} G^f(u,t) = -f\left(\lambda\left(I-B\right)\right) G^f(u,t) \\ G^f(u,0) = 1 \end{cases} \tag{5.48}$$

and its solution reads

$$\mathbb{E}\left[u^{N^f(t)}\right] = e^{-tf(\lambda(1-u))} \tag{5.49}$$

In view of (5.46)

$$\begin{aligned} \mathbb{E}[u^{N(H^f(t))}] &= \mathbb{E}\left[\mathbb{E}\left[u^{N(H^f(t))}\right] \,\middle|\, H^f(t)\right] = \\ &= \mathbb{E}[e^{-\lambda(1-u)\cdot H^f(t)}] = e^{-f(\lambda(1-u))t} \end{aligned} \tag{5.50}$$

Space-fractional Poisson process. If $f(x) = x^\alpha$, with $0 < \alpha < 1$, then the subordinator $H^f(t)$ has Laplace transform

$$\begin{aligned} &\mathbb{E}[e^{-\mu\cdot H^f(t)}] = \\ &= \exp\left\{-t\int_0^\infty (1-e^{-s\mu})\cdot\frac{\alpha s^{-\alpha-1}}{\Gamma(1-\alpha)}ds\right\} = \\ &= \exp\left\{-t\mu\int_0^\infty\left(\int_0^s e^{-\mu w}dw\right)\cdot\frac{\alpha s^{-\alpha-1}}{\Gamma(1-\alpha)}ds\right\} = \\ &= \exp\left\{-t\mu\int_0^\infty e^{-\mu w}dw\int_w^\infty\frac{\alpha s^{-\alpha-1}}{\Gamma(1-\alpha)}ds\right\} = \\ &= \exp\left\{-\frac{t\mu}{\Gamma(1-\alpha)}\int_0^\infty e^{-w\mu}[-s^{-\alpha}]_w^\infty\right\} = \\ &= \exp\left\{-\frac{t\mu}{\Gamma(1-\alpha)}\int_0^\infty w^{1-\alpha-1}e^{-w\mu}dw\right\} = \\ &= \exp\left\{-\frac{t\mu}{\Gamma(1-\alpha)}\right\}\frac{\Gamma(1-\alpha)}{\mu^{1-\alpha}} = \\ &= e^{-\mu^\alpha t} = e^{-t\cdot f(\mu)} \end{aligned} \tag{5.51}$$

Tempered Poisson process. From equation (5.50), with $f(x) = (x+\theta)^\alpha - \theta^\alpha$ ($\theta > 0, 0 < \alpha < 1$) and $\nu(ds) = \frac{\alpha s^{-(\alpha+1)} e^{-\theta s}}{\Gamma(1-\alpha)} ds$, we have that

$$
\begin{aligned}
G^{\alpha,\theta}(u,t) &= \\
= \exp\left\{ \{-t \int_0^\infty \left(1 - e^{-\lambda(1-u)s}\right) \cdot \frac{\alpha s^{-(\alpha+1)} e^{-\theta s}}{\Gamma(1-\alpha)} ds\} \right\} &= \\
= \exp\{-t[\theta + \lambda(1-u)]^\alpha - \theta^\alpha\} &= \\
= e^{\theta^\alpha t} \sum_{k=0}^\infty \frac{[-t(\theta + \lambda(1-u))^\alpha]^k}{k!} &= \\
= e^{\theta^\alpha t} \sum_{k=0}^\infty \frac{[-t(\theta+\lambda)^\alpha]^k}{k!} \left(1 - \frac{\lambda u}{\theta + \lambda}\right)^{\alpha k} &= \\
= e^{\theta^\alpha t} \sum_{k=0}^\infty \frac{(-t(\theta+\lambda))^{\alpha k}}{k!} \cdot \sum_{m=0}^\infty \frac{\Gamma(\alpha k + 1)}{m! \Gamma(\alpha k + 1 - m)} \left(-\frac{\lambda u}{\theta+\lambda}\right)^m &= \\
= \sum_{m=0}^\infty u^m \cdot \left[\frac{(-1)^m}{m!} \frac{\lambda^m e^{\theta^\alpha t}}{(\theta+\lambda)^m} \sum_{k=0}^\infty \frac{(-t(\theta+\lambda))^{\alpha k}}{k!} \frac{\Gamma(\alpha k+1)}{\Gamma(\alpha k + 1 - m)} \right]
\end{aligned}
\tag{5.52}
$$

We obtain the Poisson process with tempered or relativistic stable subordinator: the coefficient of the p.g.f. provides its probability law, i.e.

$$
\begin{aligned}
&Pr\left\{N^{\alpha,\theta}(t) = m\right\} = \\
&= \frac{(-1)^m}{m!} \left(\frac{\lambda}{\lambda+\theta}\right)^m e^{\theta^\alpha t} \times \\
&\times \sum_{k=0}^\infty \frac{(-t(\lambda+\theta))^k}{k!} \frac{\Gamma(\alpha k+1)}{\Gamma(\alpha k + 1 - m)}, \quad m \geq 0
\end{aligned}
\tag{5.53}
$$

If $\theta = 0$ then (5.53) is equal to the space-fractional Poisson process (5.22), while if $\theta = 0$ and $\alpha = 1$ (5.53) coincides with the distribution of the homogeneous Poisson process in (1.4).

Note that

$$\begin{aligned}
\mathbb{E}\left[N^{\alpha,\theta}(t)\right] &= \alpha\lambda\theta^{\alpha-1}t \\
\mathbb{V}ar\left[N^{\alpha,\theta}(t)\right] &= \alpha\lambda\theta^{\alpha-2}(\lambda(1-\alpha)+\theta)t \\
Cov\left[N^{\alpha,\theta}(t), N^{\alpha,\theta}(s)\right] &= \alpha\lambda\theta^{\alpha-2}(\lambda(1-\alpha)+\theta)(s\wedge t)
\end{aligned} \tag{5.54}$$

Remark 5.2.
To show that $N^{1,0}(t) \stackrel{d}{=} N(t)$, substitute $\theta = 0$, $\alpha = 1$ in (5.53)

$$\lim_{\substack{\alpha\to 1\\ \theta\to 0}} \Pr\{N^{\alpha,\theta}(t) = k\} =$$

$$= (-1)^k \sum_{m=0}^{\infty} \frac{(-\lambda t)^m}{m!} \frac{m!}{k!(m-k)!} =$$

$$= (-1)^k \frac{(-\lambda t)^k}{k!} \sum_{h=0}^{\infty} \frac{(-\lambda t)^h}{h!} = e^{-\lambda t} \frac{(\lambda t)^k}{k!}.$$

Negative Binomial process. If $f(x) = \log(1+x)$ and the related Lévy measure is $\nu(ds) = \frac{e^{-s}}{s} ds, \quad s > 0$, we have a Poisson process $N^{\Gamma}(t)$, $t > 0$, with the Gamma-subordinator. Its probability generating function is

$$G^{\Gamma}(u,t) = \exp\left\{-t \cdot \log(1 + \lambda(1-u))\right\} = \frac{1}{[1+\lambda(1-u)]^t} \tag{5.55}$$

The probability distribution of $N^{\Gamma}(t)$ reads

$$Pr\left\{N^{\Gamma}(t) = k\right\} = \frac{\lambda^k \Gamma(k+t)}{\Gamma(t) k! (\lambda+1)^{k+t}} \quad k \geq 0 \tag{5.56}$$

We can rewrite the p.g.f. $G^{\Gamma}(u,t)$ as

$$
\begin{aligned}
G^{\Gamma}(u,t) &= [1+\lambda(1-u)]^{-t} = \\
&= \left(1-\frac{\lambda u}{1+\lambda}\right)^{-t}(1+\lambda)^{-t} = \\
&\qquad \text{(binomial series)} \\
&= (1+\lambda)^{-t}\cdot\sum_{k=0}^{\infty}\frac{\Gamma(1-t)}{k!\Gamma(1-t-k)}\cdot\left(\frac{-\lambda u}{1+\lambda}\right)^{k} = \\
&\qquad \text{(Gamma reflection)} \\
&= (1+\lambda)^{-t}\cdot\sum_{k=0}^{\infty}\frac{1}{k!}\frac{\frac{1}{\Gamma(t)}\frac{\pi}{\sin(\pi t)}}{\frac{1}{\Gamma(k+t)}\frac{\pi}{\sin(\pi(k+t))}}\cdot\left(\frac{-\lambda u}{1+\lambda}\right)^{k} = \\
&= (1+\lambda)^{-t}\cdot\sum_{k=0}^{\infty}\frac{\Gamma(k+t)(-1)^k\sin(\pi t)}{k!\Gamma(t)\sin(\pi t)}\cdot\left(\frac{-\lambda u}{1+\lambda}\right)^{k} = \\
&= (1+\lambda)^{-t}\sum_{k=0}^{\infty}\frac{\Gamma(t+k)}{k!\Gamma(t)}\frac{\lambda^k u^k}{(1+\lambda)^k} = \\
&= \sum_{k=0}^{\infty}\frac{\Gamma(t+k)}{k!\Gamma(t)}\frac{\lambda^k}{(1+\lambda)^{t+k}}\cdot u^k
\end{aligned}
\tag{5.57}
$$

and hence (5.56) follows immediately.

Note that the intermediate step is due to the *Euler's reflection formula* of the Gamma function (see Appendix D).

It has the form of a Negative Binomial distribution, i.e.

$$
Pr\left\{B^i=k\right\} = \frac{\Gamma(i+k)}{\Gamma(i)\Gamma(k+1)}p^i q^k = \binom{k+i-1}{k}p^i q^k, \quad k\geq 0, i\geq 1.
\tag{5.58}
$$

With $p=(1+\lambda)^{-1}$, $q=\lambda(1+\lambda)^{-1}$ and $i=t$ then distribution (5.58) coincides with (5.56).

The distribution of jumps (*local* jump distribution) in this case is

$$Pr\left\{N^{\Gamma}(t,t+dt]=k\right\}=\begin{cases}\left(\frac{\lambda}{\lambda+1}\right)^{k}\frac{1}{k}dt & k\geq 1\\ 1-\log(1+\lambda)dt & k=0\end{cases} \tag{5.59}$$

which is a "Logarithmic distribution".

From (5.45) we have that

$$\begin{aligned}\Pr\{N^{\Gamma}(t.t+dt]=k\}&=\left(\frac{\lambda^{k}}{k!}\int_{0}^{\infty}e^{-\lambda s}s^{k}\frac{e^{-s}}{s}ds\right)dt=\\&=\frac{1}{k}\left(\frac{\lambda}{1+\lambda}\right)^{k}dt,\qquad k\geq 1\end{aligned} \tag{5.60}$$

From the logarithmic expansion

$$\log(1-x)=\sum_{k=1}^{\infty}(-1)^{k+1}\frac{(-x)^{k}}{k}=-\sum_{k=1}^{\infty}\frac{x^{k}}{k}\qquad 0<x<1 \tag{5.61}$$

we have that

$$\begin{aligned}\sum_{k=1}^{\infty}\Pr\{N^{\Gamma}(t,t+dt)=k\}&=\sum_{k=1}^{\infty}\frac{1}{k}\left(\frac{\lambda}{1+\lambda}\right)^{k}dt=\\&=-\log\left(1-\frac{\lambda}{1+\lambda}\right)dt=\\&=-\log\left(\frac{1}{1+\lambda}\right)dt=\log\left(1+\lambda\right)dt\end{aligned} \tag{5.62}$$

Thus, for $k=0$

$$\Pr\{N^{\Gamma}(t,t+dt)=0\}=1-\log(1+\lambda)dt \tag{5.63}$$

The main moments of $N^{\Gamma}(t)$ are

$$\begin{aligned}\mathbb{E}\left[N^{\Gamma}(t)\right]&=\lambda t\\ \mathbb{V}ar\left[N^{\Gamma}(t)\right]&=\lambda(\lambda+1)t\\ Cov\left[N^{\Gamma}(s),N^{\Gamma}(t)\right]&=\lambda(\lambda+1)(s\wedge t).\end{aligned} \tag{5.64}$$

From the probability generating function of the Negative Binomial process

$$G^{\Gamma}(u,t) = (1+\lambda(1-u))^{-t} \tag{5.65}$$

we have that

$$\mathbb{E}[N^{\Gamma}(t)] = \frac{d}{du}G^{\Gamma}(u,t)\mid_{u=1} = -t[1+\lambda(1-u)]^{-t-1}(-\lambda)\mid_{u=1} = \lambda t \tag{5.66}$$

$$\begin{aligned}\mathbb{E}[N^{\Gamma}(t)(N^{\Gamma}(t)-1)] &= \frac{d^2}{du^2}G^{\Gamma}(u,t)\mid_{u=1}= \\ \to \lambda^2 t(t+1)[1+\lambda(1-u)]^{-t-2}&\mid_{u=1}= \\ &= \lambda^2 t(t+1)\end{aligned} \tag{5.67}$$

and thus

$$\begin{aligned}\mathbb{V}ar[N^{\Gamma}(t)] &= \mathbb{E}\left[N^{\Gamma}(t)(N^{\Gamma}(t)-1)\right] + \mathbb{E}\left[N^{\Gamma}(t)\right] - \left(\mathbb{E}[N^{\Gamma}(t)]\right)^2 = \\ &= \lambda^2 t(t+1) + \lambda t - (\lambda t)^2 = \lambda^2 t + \lambda t = \lambda(\lambda+1)t\end{aligned} \tag{5.68}$$

By similar calculations, we have that

$$\mathbb{C}ov[N^{\Gamma}(t), N^{\Gamma}(t)] = \lambda(\lambda+1)\min(s,t) \tag{5.69}$$

An extension of the Binomial relation. Note that

$$Pr\left\{N^{\Gamma}(s) = h | N^{\Gamma}(t) = k\right\} = \binom{k}{h}\frac{B(s+h, t-s+k-h)}{B(s,t-s)} \qquad k \geq h, t \geq s \tag{5.70}$$

Indeed, for $s < t$, it holds true that

$$\begin{aligned}
\Pr\{N^\Gamma(s) = h \mid N^\Gamma(t) = k\} &= \frac{\Pr\{N^\Gamma(s) = h, N^\Gamma(t) = k\}}{\Pr\{N^\Gamma(t) = k\}} = \\
&= \frac{\Pr\{N^\Gamma(s) = h\} \cdot \Pr\{N^\Gamma(t) - N^\Gamma(s) = k - h\}}{\Pr\{N^\Gamma(t) = k\}} = \\
&= \frac{\frac{\lambda^h \Gamma(h+s)}{(1+\lambda)^{h+s} h! \Gamma(s)} \cdot \frac{\lambda^{k-h}}{(1+\lambda)^{k-h+t-s}} \frac{\Gamma(k-h+t-s)}{(k-h)! \Gamma(t-s)}}{\frac{\lambda^k}{(1+\lambda)^{k+t}} \frac{\Gamma(k+t)}{k! \Gamma(t)}} = \\
&= \frac{k!}{h!(k-h)!} \frac{\Gamma(t)}{\Gamma(s)\Gamma(t-s)} \frac{\Gamma(k-h+t-s)\Gamma(h+s)}{\Gamma(k+t)} = \\
&= \binom{k}{h} \frac{\Gamma(t)}{\Gamma(s)\Gamma(t-s)} \frac{\Gamma(k-h+t-s)\Gamma(h+s)}{\Gamma(k+t)} = \\
&= \binom{k}{h} \frac{B(h+s, k-h+t-s)}{B(s, t-s)}
\end{aligned} \tag{5.71}$$

where

$$\binom{k}{h} \frac{B(h+s, k-h+t-s)}{B(s, t-s)} = \binom{k}{h} \frac{\int_0^1 x^{h+s-1}(1-x)^{k-h+t-s-1} dx}{B(s, t-s)} \tag{5.72}$$

The result can also be formulated as

$$Pr\left\{N^\Gamma(s) = h | N^\Gamma(t) = k\right\} = \mathbb{E}\left[\binom{k}{h} X^h (1-X)^{k-h}\right] \qquad k \geq h, t \geq s \tag{5.73}$$

where X is a Beta random variable with parameters s and $t - s$, that is

$$Pr\left\{X \in dx\right\} = \frac{x^{s-1}(1-x)^{t-s-1}}{B(s, t-s)} dx, \quad 0 < x < 1 \tag{5.74}$$

with $B(p, q) = \int_0^1 x^{p-1}(1-x)^{q-1} dx = \frac{\Gamma(p)\Gamma(q)}{\Gamma(p+q)}$.

In fact,

$$\Pr\{N^\Gamma(s) = h \mid N^\Gamma(t) = k\} = \mathbb{E}\left[\binom{k}{h} X^h (1-X)^{k-h}\right] =$$

$$= \binom{k}{h} \int_0^1 x^h (1-x)^{k-h} \cdot \frac{x^{s-1}(1-x)^{t-s-1}}{B(s,t-s)} dx = \tag{5.75}$$

$$= \binom{k}{h} \frac{\int_0^1 x^{h+s-1}(1-x)^{k-h+t-s-1} dx}{B(s,t-s)}$$

Other properties. The Γ-Poisson process has the structure of a renewal process with intertime U possessing distribution

$$Pr\{U > t\} = G^\Gamma(0,t) = (1+\lambda)^{-t} \tag{5.76}$$

The process $N^\Gamma(t) \overset{D}{\sim} N[H^\Gamma(t)]$ has Laplace transform

$$\mathbb{E}\left[e^{-\mu H^f(t)}\right] = (1+\mu)^{-t} \tag{5.77}$$

Remark 5.3.
Note that the result generalizes the property of the Poisson process reported in equation (1.51).

5.3 Another Fractional Generalization of the Poisson Process

An alternative fractional Poisson process $\hat{N}_\nu(t)$ can be constructed by replacing the factorial function in the Poisson distribution with the Gamma function and writing the probabilities as follows (see Beghin and Orsingher, 2009)

$$Pr\left\{\hat{N}_\nu(t) = k\right\} = \frac{1}{E_{\nu,1}(\lambda t)} \frac{(\lambda t)^k}{\Gamma(\nu k + 1)}, \qquad k \geq 0 \tag{5.78}$$

where

$$E_{\nu,1}(x) = \sum_{k=0}^{\infty} \frac{x^k}{\Gamma(\nu k + 1)} \tag{5.79}$$

is the Mittag-Leffler function, with $\nu, \mu > 0$

Theorem 5.4. *The probability generating function of* (5.78) *is*

$$G_\nu(u,t) = \mathbb{E}\left[u^{\hat{N}_\nu(t)}\right] = \frac{E_{\nu,1}(\lambda ut)}{E_{\nu,1}(\lambda t)} \qquad 0 \le u \le 1 \tag{5.80}$$

which satisfies an order-ν fractional differential equation in u (see Appendix C):

$$\begin{cases} \frac{\partial^\nu G_\nu(u^\nu,t)}{\partial u^\nu} = \lambda t G_\nu(u^\nu,t) & 0<\nu<1 \\ G_\nu(u,0)=1 \end{cases} \tag{5.81}$$

Proof. The derivation reads as follows:

$$\begin{aligned}
&\frac{\partial^\nu G_\nu(u^\nu t)}{\partial u^\nu} = \\
&= \frac{1}{E_{\nu,1}(\lambda t)}\cdot\frac{1}{\Gamma(1-\nu)}\int_0^u \frac{d}{dv}\left[E_{\nu,1}(\lambda v^\nu t)\right](u-v)^{-\nu}dv = \\
&= \frac{1}{E_{\nu,1}(\lambda t)}\cdot\frac{1}{\Gamma(1-\nu)}\int_0^u \frac{d}{dv}\left[\sum_{k=0}^{\infty}\frac{(\lambda t v^\nu)^k}{\Gamma(\nu k+1)}\right](u-v)^{-\nu}dv = \\
&= \frac{1}{E_{\nu,1}(\lambda t)}\cdot\sum_{k=0}^{\infty}\frac{(\lambda t)^k}{\Gamma(\nu k+1)}\cdot\frac{1}{\Gamma(1-\nu)}\int_0^u \frac{d}{dv}\left[v^{\nu k}\right](u-v)^{-\nu}dv = \\
&\qquad \text{(by setting } v = uw\text{)} \\
&= \frac{1}{E_{\nu,1}(\lambda t)}\sum_{k=1}^{\infty}\frac{(\lambda t)^k(\nu k)u^{\nu k-1-\nu+1}}{\Gamma(\nu k+1)\Gamma(1-\nu)}\underbrace{\int_0^1 w^{\nu k-1}(1-w)^{1-\nu-1}dw}_{B(\nu k,1-\nu)} = \\
&= \frac{1}{E_{\nu,1}(\lambda t)}\sum_{k=1}^{\infty}\frac{(\lambda t)^k(\nu k)u^{\nu k-\nu}}{\Gamma(\nu k+1)\Gamma(1-\nu)}\frac{\Gamma(\nu k)\Gamma(1-\nu)}{\Gamma(\nu k+1-\nu)} = \\
&\qquad \text{(by setting } k-1=k'\text{)} \\
&= \frac{1}{E_{\nu,1}(\lambda t)}\sum_{k'=0}^{\infty}\frac{(\lambda t)^{k'+1}u^{\nu k'}}{\Gamma(\nu k'+1)} = \lambda t\frac{E_{\nu,1}(\lambda u^\nu t)}{E_{\nu,1}(\lambda t)} = \\
&= \lambda t\cdot G_\nu(u^\nu,t)
\end{aligned} \tag{5.82}$$

□

The moments. Given the probability generating function of the model (5.80), one obtains the first and second moments from the corresponding first and second derivatives as

$$\begin{cases} \frac{d}{du} G^{\nu}(u,t) \mid_{u=1} = \mathbb{E}[\hat{N}(t)] \\ \frac{d^2}{du^2} G^{\nu}(u,t) \mid_{u=1} = \mathbb{E}[\hat{N}(t)(\hat{N}(t)-1)] \end{cases} \tag{5.83}$$

From the definition of the Mittag-Leffler function $E_{\nu,1}(\lambda tu) = \sum_{k=0}^{\infty} \frac{(\lambda tu)^k}{\Gamma(\nu k+1)}$, the first and second derivatives follow immediately.

The expected value of $\hat{N}_\nu(t)$ is

$$\mathbb{E}\left[\hat{N}_\nu(t)\right] = \frac{\lambda t}{\nu} \frac{E_{\nu,\nu}(\lambda ut)}{E_{\nu,1}(\lambda t)} \mid_{u=1} = \frac{\lambda t}{\nu} \frac{E_{\nu,\nu}(\lambda t)}{E_{\nu,1}(\lambda t)} \tag{5.84}$$

In fact, the first derivative is:

$$\begin{aligned} \frac{d}{du} E_{\nu,1}(\lambda tu) = \sum_{k=1}^{\infty} \frac{(\lambda t)^k k u^{k-1}}{\nu k \Gamma(\nu k)} &= (k-1=k') = \\ = \frac{1}{\nu} \sum_{k'=0}^{\infty} \frac{(\lambda t)^{k'+1} u^{k'}}{\Gamma(\nu(k'+1))} &\equiv \\ \equiv \frac{\lambda t}{\nu} \sum_{k=0}^{\infty} \frac{(\lambda t)^k u^k}{\Gamma(\nu k+\nu)} = \frac{\lambda t}{\nu} E_{\nu,\nu}(\lambda tu) & \end{aligned} \tag{5.85}$$

and the mean obtains as (for $u = 1$)

$$\mathbb{E}[\hat{N}(t)] = \frac{\lambda t}{\nu} \frac{E_{\nu,\nu}(\lambda t)}{E_{\nu,1}(\lambda t)} \tag{5.86}$$

While the variance is:

$$\mathbb{V}ar\left[\hat{N}_\nu(t)\right] = \frac{\lambda t}{\nu^2} \frac{E_{\nu,\nu-1}(\lambda t)}{E_{\nu,1}(\lambda t)} + \frac{\lambda t}{\nu^2} \frac{E_{\nu,\nu}(\lambda t)}{E_{\nu,1}(\lambda t)} \left\{1 - \lambda t \frac{E_{\nu,\nu}(\lambda t)}{E_{\nu,1}(\lambda t)}\right\} \tag{5.87}$$

For the second derivative, the following computations hold:

$$
\begin{aligned}
\frac{d^2}{du^2}E_{\nu,1}(\lambda tu) = & \\
& = \sum_{k=2}^{\infty} \frac{(\lambda t)^k k(k-1)u^{k-2}}{\nu k\Gamma(\nu k)} = \\
& (k-1=k') \\
& = \frac{1}{\nu}\sum_{k'=1}^{\infty} \frac{(\lambda t)^{k'+1} k' u^{k'-1}}{\Gamma(\nu(k'+1))} = \\
& = \frac{\lambda t}{\nu}\sum_{k=0}^{\infty} \frac{(\lambda t)^k k u^{k-1}}{\Gamma(\nu(k+1)+1-1)} = \\
& = \frac{\lambda t}{\nu}\sum_{k=0}^{\infty} \frac{(\lambda t)^k k u^{k-1}}{(\nu(k+1)-1)\Gamma(\nu(k+1)-1)} = \\
& = \frac{\lambda t}{\nu^2}\sum_{k=0}^{\infty} \frac{(\lambda t)^k (\nu k+\nu-1-(\nu-1)) u^{k-1}}{(\nu(k+1)-1)\Gamma(\nu(k+1)-1)} = \\
& = \frac{\lambda t}{\nu^2}\left[\sum_{k=0}^{\infty} \frac{(\lambda t)^k u^{k-1}}{\Gamma(\nu(k+1)-1)} + \right. \\
& \left. -(\nu-1)\sum_{k=0}^{\infty} \frac{(\lambda t)^k u^{k-1}}{\Gamma(\nu(k+1))}\right]
\end{aligned}
\tag{5.88}
$$

that is

$$
\frac{d^2}{du^2}E_{\nu,1}(\lambda tu) \,|_{u=1} = \frac{\lambda t}{\nu^2}E_{\nu,\nu-1}(\lambda t) - (\nu-1)\frac{\lambda t}{\nu^2}E_{\nu,\nu}(\lambda t) \tag{5.89}
$$

Since

$$
\mathbb{V}ar[\hat{N}(t)] = \mathbb{E}[\hat{N}(t)(\hat{N}(t)-1)] + \mathbb{E}[\hat{N}(t)] - \Big(\mathbb{E}[\hat{N}(t)]\Big)^2 \tag{5.90}
$$

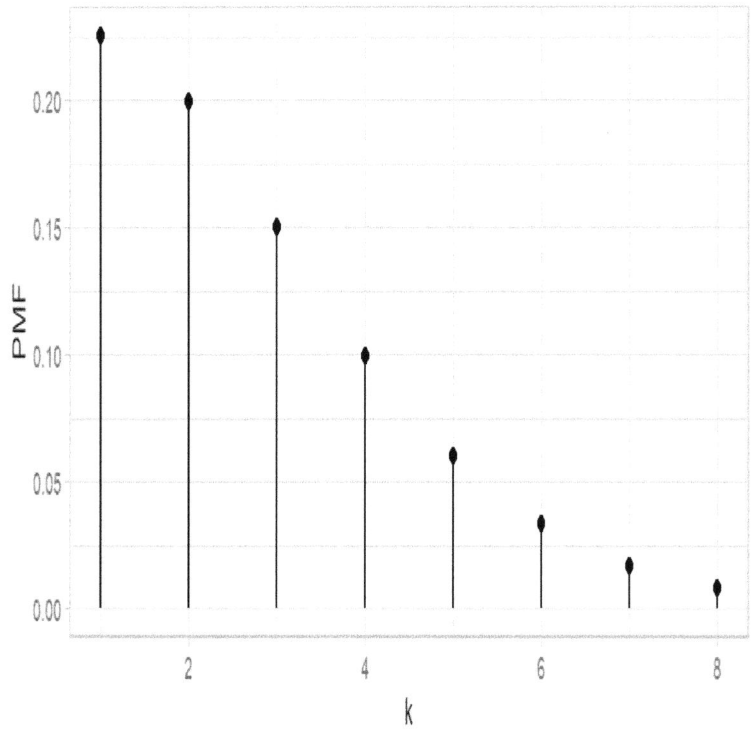

FIGURE 5.4
Distribution $p_k(t), k \geq 1$ for $\lambda = 1$ and $\nu = 0.5$.

one obtains $(u = 1)$

$$\mathbb{V}ar[\hat{N}(t)] =$$

$$= \frac{1}{E_{\nu,1}(\lambda t)} \cdot \left\{ \frac{\lambda t}{\nu^2} E_{\nu,\nu-1}(\lambda t) + \frac{\lambda t}{\nu^2} E_{\nu,\nu}(\lambda t) \right\} - \left(\frac{\lambda t}{\nu} \frac{E_{\nu,\nu}(\lambda t)}{E_{\nu,1}(\lambda t)} \right)^2 \tag{5.91}$$

For $\nu = 1$, because $E_{1,1}(x) = e^x$ and $E_{1,0}(x) = xe^x$, then $\hat{N}_1(t)$ is the simple Poisson process. For $\nu < 1$, $\hat{N}_\nu(t)$ has no longer independent increments.

It is easy to show that for i.i.d. r.v.'s with distribution function $F(x)$:

$$Pr\left\{ \max\left(X_1, \dots, X_{\hat{N}_\nu(t)} \right) < u \right\} = \frac{E_{\nu,1}(\lambda t F(u))}{E_{\nu,1}(\lambda t)}. \tag{5.92}$$

In fact

$$\begin{aligned}
&\Pr\{\max\{X_1, ..., X_{\hat{N}_\nu(t)}\} < u\} = \\
&= \mathbb{E}_{\hat{N}_\nu}[\Pr\{\max\{X_1, ..., X_{\hat{N}_\nu(t)}\} < u|\hat{N}_\nu(t)\}] = \\
&= \mathbb{E}_{\hat{N}_\nu}[\Pr(X_1 < u)^{\hat{N}_\nu(t)}] = \\
&= \sum_{k=0}^{\infty} \Pr(X_1 < u)^k \frac{(\lambda t)^k}{\Gamma(\nu k + 1)} \frac{1}{E_{\nu,1}(\lambda t)} = \\
&= \frac{E_{\nu,1}(\lambda t F(u))}{E_{\nu,1}(\lambda t)}.
\end{aligned} \tag{5.93}$$

Moreover,

$$\Pr\{\min(X_1, ..., X_{\hat{N}_\nu(t)}) > u\} = \frac{E_{\nu,1}(\lambda t(1 - F(u)))}{E_{\nu,1}(\lambda t)} \tag{5.94}$$

The distribution of the waiting time of the first event T_1^ν is

$$\Pr\{\hat{T}_1^\nu > t\} = \Pr(\hat{N}_\nu(t) = 0) = \frac{1}{E_{\nu,1}(\lambda t)}. \tag{5.95}$$

Remark 5.4.
The distribution (5.78) can be seen as a weighted transformation of the homogeneous Poisson process (see Fig. 5.4):

$$\Pr\left\{\hat{N}_\nu(t) = k\right\} = \frac{Pr\{N(t) = k\}\frac{k!}{\Gamma(\nu k+1)}}{\sum_{j=0}^{\infty} Pr\{N(t) = j\}\frac{j!}{\Gamma(\nu j+1)}} \qquad k \geq 0$$

We can verify this result by substitution, i.e.,

$$\begin{aligned}
&\Pr\left\{\hat{N}_\nu(t) = k\right\} = \\
&= \frac{\Pr\{N(t) = k\} \cdot \frac{k!}{\Gamma(\nu k+1)}}{\sum_{h=0}^{\infty} \Pr\{N(t) = h\} \cdot \frac{h!}{\Gamma(\nu h+1)}} = \\
&= \frac{e^{-\lambda t}\frac{(\lambda t)^k}{k!} \cdot \frac{k!}{\Gamma(\nu k+1)}}{\sum_{h=0}^{\infty} e^{-\lambda t}\frac{(\lambda t)^h}{h!} \cdot \frac{h!}{\Gamma(\nu h+1)}} = \\
&= \frac{\frac{k!}{\Gamma(\nu k+1)}}{\sum_{h=0}^{\infty} \frac{(\lambda t)^h}{\Gamma(\nu h+1)}} = \frac{1}{E_{\nu,1}(\lambda t)} \cdot \frac{(\lambda t)^k}{\Gamma(\nu k + 1)}
\end{aligned} \tag{5.96}$$

5.4 The Time-fractional Poisson Process

In this case it is useful to start the analysis from the equation governing the state probabilities (see Beghin and Orsingher, 2009) generalized by a (time)-fractional derivative

$$\frac{d^\nu p_k(t)}{dt^\nu} = -\lambda p_k(t) + \lambda p_{k-1}(t) \quad k \geq 0, p_{-1} = 0, 0 < \nu < 1 \tag{5.97}$$

subject to the initial conditions

$$p_k(0) = \begin{cases} 1 & k = 0, \\ 0 & k \geq 1, \end{cases} \tag{5.98}$$

where

$$\frac{d^\nu p_k(t)}{dt^\nu} = \frac{1}{\Gamma(1-\nu)} \int_0^t \frac{dp_k(s)}{ds} \frac{1}{(t-s)^\nu} ds \tag{5.99}$$

is the fractional derivative in the sense of Dzherbashyan-Caputo.

Theorem 5.5. *The p.g.f. $G_\nu(u,t)$ of the time-fractional Poisson process solves the Cauchy problem*

$$\begin{cases} \frac{\partial^\nu}{\partial t^\nu} G_\nu(u,t) = -\lambda(1-u) G_\nu(u,t) & 0 < \nu < 1 \\ G_\nu(u,0) = 1 \end{cases} \tag{5.100}$$

and has the form

$$G_\nu(u,t) = E_{\nu,1}\left(-\lambda(1-u)t^\nu\right) \tag{5.101}$$

Proof. By taking the Laplace transform of both members of (5.100) we have

$$\int_0^\infty e^{-\mu t} \frac{\partial^\nu}{\partial t^\nu} G_\nu(u,t) dt = \\ = -\lambda(1-u) \int_0^\infty e^{-\mu t} G_\nu(u,t) dt \tag{5.102}$$

From (5.102) and taking into account the initial condition we have that

$$\begin{aligned} \mu^{\nu}\int_0^{\infty} e^{-\mu t}G_{\nu}(u,t)dt - \mu^{\nu-1}G_{\nu}(u,0) = \\ = -\lambda(1-u)\int_0^{\infty} e^{-\mu t}G_{\nu}(u,t)dt \end{aligned} \tag{5.103}$$

and thus

$$\int_0^{\infty} e^{-\mu t}G_{\nu}(u,t)dt = \frac{\mu^{\nu-1}}{\mu^{\nu}+\lambda(1-u)} \tag{5.104}$$

The inverse Laplace transform is

$$G_{\nu}(u,t) = E_{\nu,1}\left(-\lambda(1-u)t^{\nu}\right) \tag{5.105}$$

because

$$\begin{aligned} &\int_0^{\infty} e^{-\mu t}G_{\nu}(u,t)dt = \\ &= \int_0^{\infty} e^{-\mu t}E_{\nu,1}\left(-\lambda(1-u)t^{\nu}\right)dt = \\ &= \sum_{k=0}^{\infty}\frac{(-\lambda(1-u))^k}{\Gamma(\nu k+1)}\cdot\int_0^{\infty} e^{-\mu t}t^{\nu k}dt = \\ &= \sum_{k=0}^{\infty}\frac{(-\lambda(1-u))^k}{\Gamma(\nu k+1)}\frac{\Gamma(\nu k+1)}{\mu^{\nu k+1}} = \\ &= \frac{1}{\mu}\frac{1}{1+\frac{\lambda(1-u)}{\mu^{\nu}}} = \frac{\mu^{\nu-1}}{\mu^{\nu}+\lambda(1-u)} \end{aligned} \tag{5.106}$$

□

For $\nu = 1$ (5.101) coincides with the p.g.f. of the Poisson process, i.e.

$$G_1(u,t) = E_{1,1}(-\lambda(1-u)t) = e^{-\lambda(1-u)t} \tag{5.107}$$

It can be shown (see Beghin and Orsingher, 2009) that:

Theorem 5.6. *The distribution of the time-fractional Poisson process from* (5.97) *has the form*

$$\begin{aligned} p_k^\nu(t) &= \frac{(\lambda t^\nu)^k}{k!}\sum_{h=0}^{\infty}\frac{(h+k)!}{h!}\frac{(-\lambda t^\nu)^h}{\Gamma(\nu(k+h)+1)} \\ &= \sum_{h=k}^{\infty}(-1)^{h-k}\binom{h}{k}\frac{(\lambda t^\nu)^h}{\Gamma(\nu h+1)} \qquad k \geq 0 \end{aligned} \tag{5.108}$$

Proof. The probability generating function of the time-fractional Poisson process is

$$\begin{aligned} G_\nu(u,t) = \mathbb{E}\left[u^{N_\nu(t)}\right] &= \\ = E_{\nu,1}(\lambda(u-1)t^\nu) = \sum_{h=0}^{\infty}\frac{(\lambda(u-1)t^\nu)^h}{\Gamma(\nu h+1)} &= \\ = \sum_{h=0}^{\infty}\frac{(\lambda t^\nu)^h}{\Gamma(\nu h+1)}\cdot\sum_{k=0}^{h}\binom{h}{k}u^k(-1)^{h-k} &= \\ = \sum_{k=0}^{\infty}u^k\cdot\sum_{h=k}^{\infty}\binom{h}{k}\frac{(\lambda t^\nu)^h(-1)^{h-k}}{\Gamma(\nu h+1)} & \end{aligned} \tag{5.109}$$

hence, the probability reads

$$\begin{aligned} p_k^\nu(t) = \Pr\{N_\nu(t) = k\} &= \\ = \sum_{h=k}^{\infty}(-1)^{h-k}\binom{h}{k}\frac{(\lambda t^\nu)^h}{\Gamma(\nu h+1)} &= \\ = \frac{(\lambda t^\nu)^k}{k!}\sum_{h=0}^{\infty}\frac{(h+k)!}{h!}\frac{(-\lambda t^\nu)^h}{\Gamma(\nu(k+h)+1)} & \end{aligned} \tag{5.110}$$

□

The above distribution can also be restated as

$$p_k^\nu(t) = \frac{1}{k!}\sum_{h=k}^{\infty}\frac{(-1)^{h-k}}{(h-k)!}\mathbb{E}\left[N_\nu(t)...(N_\nu(t)-h+1)\right] \tag{5.111}$$

For the probability generating function, the Maclaurin expansion yields

$$G_\nu(u,t) = \sum_{k=0}^{\infty}\frac{\partial^k G_\nu(u,t)}{\partial u^k}\mid_{u=0}\cdot\frac{u^k}{k!} \tag{5.112}$$

Since

$$\begin{aligned}\frac{\partial^k}{\partial u^k}G_\nu(u,t)\mid_{u=0} &= \frac{d^k}{du^k}\left\{\sum_{h=0}^{\infty}\frac{(\lambda t^\nu)^h(u-1)^h}{\Gamma(\nu h+1)}\right\}\mid_{u=0}= \\ =\sum_{h=0}^{\infty}\frac{(\lambda t^\nu)^h}{\Gamma(\nu h+1)}&[h(h-1)...(h-k+1)](u-1)^{h-k}\mid_{u=0}= \\ =\sum_{h=k}^{\infty}\frac{(\lambda t^\nu)^h}{\Gamma(\nu h+1)}&\frac{[h(h-1)...(h-k+1)](h-k)!}{(h-k)!}(-1)^{h-k} = \\ &= k!\sum_{h=k}^{\infty}\frac{(\lambda t^\nu)^h}{\Gamma(\nu h+1)}\binom{h}{k}(-1)^{h-k}\end{aligned} \tag{5.113}$$

Hence

$$\begin{aligned}G_\nu(u,t) &= \sum_{k=0}^{\infty}\frac{u^k}{k!}\cdot\sum_{h=k}^{\infty}\frac{k!(\lambda t^\nu)^h}{\Gamma(\nu h+1)}\binom{h}{k}(-1)^{h-k} = \\ &= \sum_{k=0}^{\infty}u^k\cdot\Pr\{N_\nu(t)=k\}\end{aligned} \tag{5.114}$$

Then it holds true that

$$
\begin{aligned}
&\frac{\partial^k}{\partial u^k} G_\nu(u,t)\mid_{u=1} = \mathbb{E}\left[N_\nu(t)...(N_\nu(t)-k+1)\right] = \\
&= \sum_{h=k}^{\infty} [h(h-1)...(h-k+1)] \cdot \Pr\{N_\nu(t)=h\} = \\
&\qquad \text{(from (5.110))} \\
&= \sum_{h=k}^{\infty} \frac{(\lambda t^\nu)^h}{\Gamma(\nu h+1)} [h(h-1)...(h-k+1)](u-1)^{h-k}\mid_{u=1} = \\
&\qquad (\neq 0 \text{ only for } h=k) \\
&= \frac{(\lambda t^\nu)^k}{\Gamma(\nu k+1)} [k(k-1)...2\cdot 1] = k! \frac{(\lambda t^\nu)^k}{\Gamma(\nu k+1)}
\end{aligned}
\tag{5.115}
$$

By replacing in (5.110) the relationship

$$
\frac{1}{h!}\mathbb{E}[N_\nu(t)(N_\nu(t)-1)...(N_\nu(t)-h+1)] = \frac{(\lambda t^\nu)^h}{\Gamma(\nu h+1)} \tag{5.116}
$$

we obtain:

$$
\begin{aligned}
&\Pr\{N_\nu(t)=k\} = \\
&= \sum_{h=k}^{\infty} \frac{h!}{k!(h-k)!}(-1)^{h-k} \frac{(\lambda t^\nu)^h}{\Gamma(\nu h+1)} = \\
&= \sum_{h=k}^{\infty} \frac{h!}{k!(h-k)!}(-1)^{h-k} \times \\
&\times \left\{ \frac{1}{h!}\mathbb{E}\left[N_\nu(t)(N_\nu(t)-1)...(N_\nu(t)-h+1)\right] \right\} = \\
&= \frac{1}{k!}\sum_{h=k}^{\infty} \frac{(-1)^{h-k}}{(h-k)!}\mathbb{E}\left[N_\nu(t)(N_\nu(t)-1)...(N_\nu(t)-h+1)\right]
\end{aligned}
\tag{5.117}
$$

that is the third representation of the distribution obtained in equation (2.10) by Orsingher and Beghin (2009).

Still an alternative derivation is possible. From (5.114) we know that

$$\Pr\{N_\nu(t)=k\} = \sum_{h=k}^{\infty} \binom{h}{k} (-1)^{h-k} \frac{(\lambda t^\nu)^h}{\Gamma(\nu h+1)} \tag{5.118}$$

Since

$$\frac{d^k}{du^k} G_\nu(u,t) \mid_{u=1} = \mathbb{E}\left[N_\nu(t)(N_\nu(t)-1)...(N_\nu(t)-k+1)\right] \tag{5.119}$$

and

$$\frac{d^k}{du^k} G_\nu(u,t) = \sum_{h=k}^{\infty} \frac{(\lambda t^\nu)^h h(h-1)...(h-k+1)(u-1)^{h-k}}{\Gamma(\nu h+1)} \tag{5.120}$$

then it holds true that

$$\begin{aligned}\frac{d^k}{du^k} G_\nu(u,t) \mid_{u=1} &= \mathbb{E}\left[N_\nu(t)(N_\nu(t)-1)...(N_\nu(t)-k+1)\right] = \\ &= \frac{(\lambda t^\nu)^k}{\Gamma(\nu k+1)} k(k-1)...2\cdot 1 = k! \frac{(\lambda t^\nu)^k}{\Gamma(\nu k+1)}\end{aligned} \tag{5.121}$$

Hence

$$\begin{aligned}&\Pr\{N_\nu(t)=k\} = \\ &= \sum_{h=k}^{\infty} \frac{h!}{k!(h-k)!} (-1)^{h-k} \frac{\mathbb{E}\left[N_\nu(t)(N_\nu(t)-1)...(N_\nu(t)-h+1)\right]}{h!} = \\ &= \frac{1}{k!} \sum_{h=k}^{\infty} \frac{(-1)^{h-k}}{(h-k)!} \mathbb{E}\left[N_\nu(t)(N_\nu(t)-1)...(N_\nu(t)-h+1)\right]\end{aligned} \tag{5.122}$$

For small values of k, it is possible to write the distribution (5.108) in terms of Mittag-Leffler functions as follows

$$\begin{cases} p_0^\nu(t) = E_{\nu,1}(-\lambda t^\nu) \\ p_1^\nu(t) = \frac{\lambda t^\nu}{\nu} E_{\nu,\nu}(-\lambda t^\nu) \\ p_2^\nu(t) = \frac{(\lambda t^\nu)^2}{2!\nu^2}\left[(1-\nu)E_{\nu,2\nu}(-\lambda t^\nu) + E_{\nu,2\nu-1}(-\lambda t^\nu)\right] \\ p_3^\nu(t) = \frac{(\lambda t^\nu)^3}{3!\nu^3}\left[2(1-\nu)\left(\frac{1}{2}-\nu\right)E_{\nu,3\nu}(-\lambda t^\nu) + \right. \\ \qquad \left. +3(1-\nu)E_{\nu,3\nu-1}(-\lambda t^\nu) + E_{\nu,3\nu-2}(-\lambda t^\nu)\right] \end{cases} \tag{5.123}$$

In particular,

$$p_0^\nu(t) = \sum_{h=0}^{\infty} \frac{(-\lambda t^\nu)^h}{\Gamma(\nu h+1)} = E_{\nu,1}(-\lambda t^\nu) \tag{5.124}$$

and

$$\begin{aligned} p_1^\nu(t) &= \\ = \sum_{h=0}^{\infty}(-1)\frac{-\lambda t^\nu}{1!}\sum_{h=0}^{\infty}\frac{(h+1)!}{h!}\frac{(-\lambda t^\nu)^h}{\Gamma(\nu(h+1))+1} &= \\ = (\lambda t^\nu)\sum_{h=0}^{\infty}\frac{(h+1)(-\lambda t^\nu)^h}{\nu(h+1)\Gamma(\nu(h+1))} &= \\ = \frac{\lambda t^\nu}{\nu}\sum_{h=0}^{\infty}\frac{(-\lambda t^\nu)^h}{\Gamma(\nu h+\nu)} &= \\ = \frac{\lambda t^\nu}{\nu}E_{\nu,\nu}(-\lambda t^\nu) \end{aligned} \tag{5.125}$$

When $\nu = 1$ the distribution coincides with that of the Poisson process, i.e.

$$\begin{aligned} p_k^1(t) &= \\ &= (-1)^k \sum_{h=k}^{\infty} \frac{h!}{k!(h-k)!}\frac{(-\lambda t)^h}{h!} = \\ &= (-1)^k \sum_{h=0}^{\infty} \frac{(-\lambda t)^{k+h}}{k!h!} = \\ &= \frac{(-\lambda t)^k}{k!}(-1)^k \sum_{h=0}^{\infty}\frac{(-\lambda t)^h}{h!} = \\ &= e^{-\lambda t}\frac{(\lambda t)^k}{k!} \end{aligned} \tag{5.126}$$

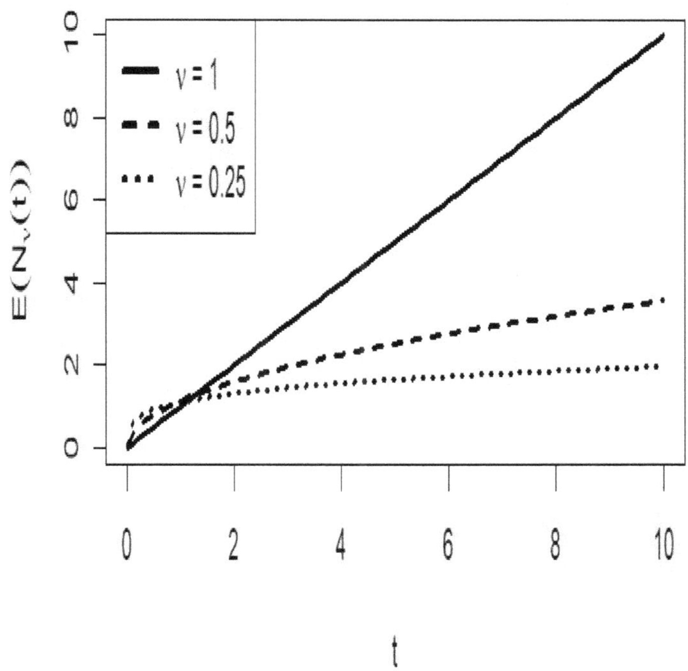

FIGURE 5.5
Comparison between expected values (w.r.t. time) of a time-fractional Poisson process of different parameters $\nu = \{1, 0.5, 0.25\}$ and $\lambda = 1$. Note: $\nu = 1$ corresponds to the standard Poisson process.

The probability generating function of the time-fractional Poisson process $N_\nu(t)$ is From (5.101) we can extract the mean and the variance (see Fig. 5.5 and Fig. 5.6), i.e.:

Theorem 5.7.

$$\begin{aligned} \mathbb{E}\left[N_\nu(t)\right] &= \frac{\lambda t^\nu}{\Gamma(\nu+1)} \\ \mathbb{V}ar\left[N_\nu(t)\right] &= \frac{\lambda t^\nu}{\Gamma(\nu+1)} + \frac{(\lambda t^\nu)^2}{\nu}\left\{\frac{1}{\Gamma(2\nu)} - \frac{1}{\nu\Gamma^2(\nu)}\right\} \end{aligned} \tag{5.127}$$

Proof. It is well known that the relation between the derivatives of a p.g.f. $G(u,t)$ (computed in $u = 1-$) and the (factorial) moments k_n of a r.v. X holds:

$$\begin{cases} G'(u,t)_{|u=1} = \mathbb{E}[X] \\ G''(u,t)_{|u=1} = \mathbb{E}[X(X-1)] \\ G^{(n)}(u,t)_{|u=1} = \mathbb{E}[X(X-1)...(X-n+1)] \end{cases} \tag{5.128}$$

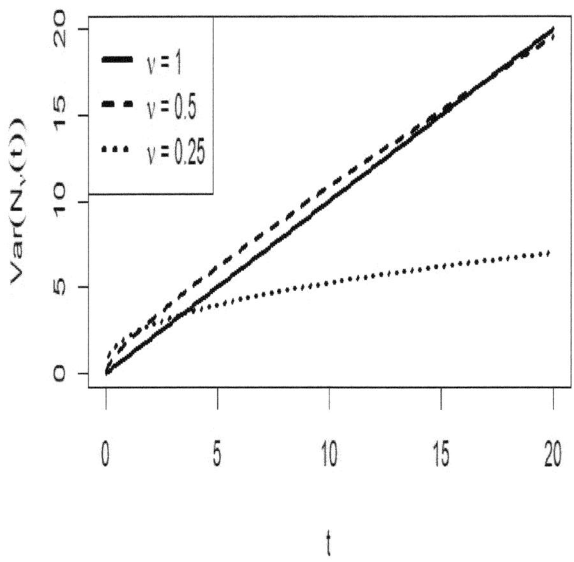

FIGURE 5.6
Comparison between variances (w.r.t. time) of a time-fractional Poisson process of different parameters, $\nu = \{1, 0.5, 0.25\}$ $(\lambda = 1)$. Note: $\nu = 1$ corresponds to the standard Poisson process.

The first derivative reads

$$\begin{aligned}
\mathbb{E}[N_\nu(t)] &= \\
&= \frac{d}{du} G-\nu(u,t)\mid_{u=1} = \\
&= \frac{d}{du} E_{\nu,1}(-\lambda(1-u)t^\nu)\mid_{u=1} = \\
&= \sum_{k=0}^{\infty} \frac{f}{du}\frac{[-\lambda(1-u)t^\nu]^k}{\Gamma(\nu k+1)}\mid_{u=1} = \\
&= -\sum_{k=0}^{\infty} \frac{(-\lambda t^\nu)^k \cdot k(1-u)^{k-1}}{\Gamma(\nu k+1)}\mid_{u=1} = \\
&= -\sum_{k=1}^{\infty} \frac{(-\lambda t^\nu)^k \nu k(1-u)^{k-1}}{\nu\Gamma(\nu k)\nu k}\mid_{u=1} = \\
&= -\sum_{k'=0}^{\infty} \frac{(-\lambda t^\nu)^{k'+1}(1-u)^{k'}}{\nu\Gamma(\nu k'+\nu)}\mid_{u=1} = \\
&= -\frac{-(\lambda t^\nu)}{\nu\Gamma(\nu)} = \frac{(\lambda t^\nu)}{\Gamma(\nu+1)}
\end{aligned} \tag{5.129}$$

As for the second derivative, it reads

$$
\begin{aligned}
\mathbb{E}[N_\nu(t)(N_\nu(t)-1)] &= \\
&= \frac{d^2}{du^2} G_\nu(u,t)_{|u=1} = \\
&= \frac{d^2}{du^2} E_{\nu,1}\left(-\lambda(1-u)t^\nu\right)_{|u=1} = \\
&= \frac{d^2}{du^2} \sum_{k=0}^{\infty} \frac{(-\lambda t^\nu)^k (1-u)^k}{\Gamma(\nu k+1)}_{|u=1} = \\
&= \sum_{k=2}^{\infty} \frac{(-\lambda t^\nu)^k \cdot k(k-1)(1-u)^{k-2}}{\Gamma(\nu k+1)}_{|u=1} = \\
&= -\frac{(-\lambda t^\nu)^2 2\cdot 1}{\Gamma(2\nu+1)} = \frac{(\lambda t^\nu)^2}{\nu\Gamma(2\nu)}
\end{aligned}
\tag{5.130}
$$

Hence, we have that the first two factorial moments are

$$
\begin{cases}
\mathbb{E}[N_\nu(t)] = \frac{\lambda t^\nu}{\Gamma(\nu+1)} \\
\mathbb{E}[N_\nu(t)(N_\nu(t)-1)] = \frac{(\lambda t^\nu)^2}{\nu\Gamma(2\nu)}
\end{cases}
\tag{5.131}
$$

Finally, from the first two factorial moments, the variance follows straightforward as

$$
\begin{aligned}
\mathbb{V}ar[N_\nu(t)] &= \mathbb{E}[N_\nu(t)(N_\nu(t)-1)] + \mathbb{E}[N_\nu(t)] - \left(\mathbb{E}[N_\nu(t)]\right)^2 = \\
&= \frac{(\lambda t^\nu)^2}{\nu\Gamma(2\nu)} + \frac{\lambda t^\nu}{\Gamma(\nu+1)} - \frac{(\lambda t^\nu)^2}{\nu^2\Gamma^2(\nu)} = \\
&= \frac{\lambda t^\nu}{\Gamma(\nu+1)} + \frac{(\lambda t^\nu)^2}{\nu}\left(\frac{1}{\Gamma(2\nu)} - \frac{1}{\nu\Gamma^2(\nu)}\right)
\end{aligned}
\tag{5.132}
$$

□

Clearly, the time-fractional Poisson process is overdispersed[3].

[3]For $0 < \nu \leq 1$, the term $\Delta(\nu) = \frac{1}{\Gamma(2\nu)} - \frac{1}{\nu\Gamma^2(\nu)} \geq 0$. In fact, by means of the Legendre duplication formula of the gamma function, i.e. $\Gamma(2z) = \frac{\Gamma(z)\Gamma(z+\frac{1}{2})}{2^{1-2z}\sqrt{\pi}}$, one can easily check that $\Delta(\nu) \geq 0$ for $0 < \nu \leq 1$.

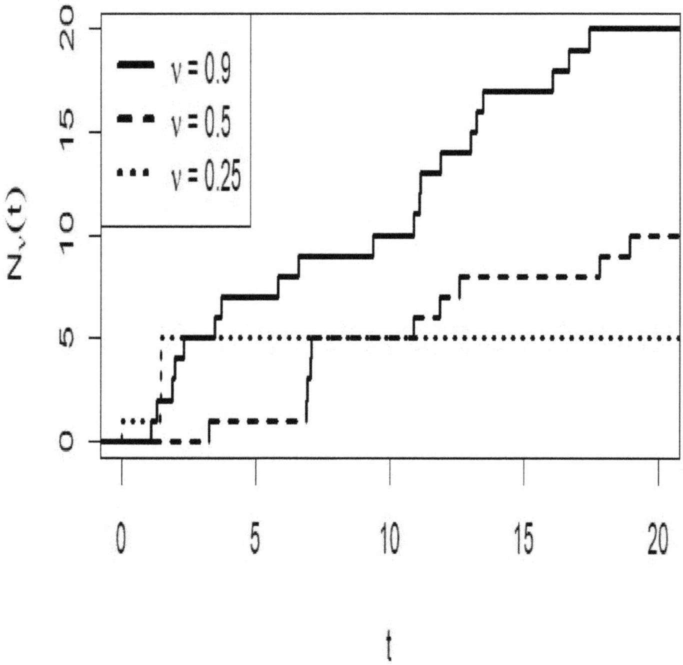

FIGURE 5.7
Time-fractional Poisson: A sample trajectory for $\nu = 0.9$, $\nu = 0.5$ and $\nu = 0.25$. Note the "slow pace" of progress of the counts w.r.t. time.

Renewal Process representation. The time-fractional Poisson process is a renewal process with intertimes $U_j = \tau_j - \tau_{j-1}$ having the distribution

$$Pr\{U_j > t\} = \Pr\{T_1^\nu > t\} = Pr\{N_\nu(t) = 0\} = E_{\nu,1}(-\lambda t^\nu) = \\ = p_0^\nu(t) \tag{5.133}$$

where $T_1^\nu(t) = \inf\{t > 0 : N_\nu(t) = 1\}$.

One can notice (see Fig. 5.7) how intertimes have longer durations than exponential, corresponding to "plateau" with no jumps, followed by clusters of (unit) jumps. Indeed, the intertimes have infinite expectations.

The density of the intertimes U_j is

$$\Pr\{U_j \in dt\} = \lambda t^{\nu-1} E_{\nu,\nu}(-\lambda t^\nu)dt, \quad t > 0, 0 < \nu < 1 \qquad (5.134)$$

In fact, by virtue of (5.133), the following computations hold:

$$\begin{aligned}
\Pr\{U_j \in dt\} &= -\frac{d}{dt}\Pr\{U_j > t\} = -\frac{d}{dt}E_{\nu,1}(-\lambda t^\nu) = \\
&= -\frac{d}{dt}\sum_{k=0}^{\infty}\frac{(-\lambda t^\nu)^k}{\Gamma(\nu k+1)} = -\sum_{k=0}^{\infty}\frac{(-\lambda)^k \frac{d}{dt}t^{\nu k}}{\Gamma(\nu k+1)} = \\
&= -\sum_{k=1}^{\infty}\frac{(-\lambda)^k \nu k t^{\nu k-1}}{\Gamma(\nu k+1)} = -\sum_{k=1}^{\infty}\frac{(-\lambda)^k t^{\nu k-1}}{\Gamma(\nu k)} = \\
&\qquad (k-1=k') \\
= -\sum_{k=0}^{\infty}\frac{(-\lambda)^{k'+1} t^{\nu(k'+1)-1}}{\Gamma(\nu k'+\nu)} &= -(-\lambda)t^{\nu-1}\sum_{k=0}^{\infty}\frac{(-\lambda)^k t^{\nu k}}{\Gamma(\nu k+\nu)} = \\
&= \lambda t^{\nu-1}\sum_{k=0}^{\infty}\frac{-(\lambda t^\nu)^k}{\Gamma(\nu k+\nu)} = \lambda t^{\nu-1}E_{\nu,\nu}(-\lambda t^\nu)
\end{aligned} \qquad (5.135)$$

Moreover, the (i.d.) renewal times U_j have a Laplace transform

$$\int_0^\infty e^{-\mu t} Pr\{U_j \in dt\} = \frac{\lambda}{\lambda+\mu^\nu}, \quad \mu > 0. \qquad (5.136)$$

In fact, by virtue of the density (5.134), the Laplace transform of the random variable U_j is obtained as

$$\begin{aligned}\int_0^\infty e^{-\mu t}\Pr\{U_j \in dt\} = \int_0^\infty e^{-mut} \cdot \lambda t^{\nu-1} E_{\nu,nu}(-\lambda t^\nu)dt = \\ = \sum_{k=0}^\infty \int_0^\infty e^{-\mu t}\lambda t^{\nu-1}\frac{(-\lambda t^\nu)^k}{\Gamma(\nu k+\nu)}dt = \\ = \sum_{k=0}^\infty \frac{\lambda(-\lambda)^k}{\Gamma(\nu k+\nu)}\int_0^\infty e^{-\mu t}t^{\nu+\nu k-1}dt = \\ = \sum_{k=0}^\infty \frac{\lambda(-\lambda)^k}{\Gamma(\nu k+\nu)}\frac{\Gamma(\nu k+\nu)}{\mu^{\nu k+\nu}} = \\ = \frac{\lambda}{\mu^\nu}\sum_{k=0}^\infty\left(-\frac{\lambda}{\mu^\nu}\right)^k = \frac{\lambda}{\mu^\nu}\frac{1}{1+\frac{\lambda}{\mu^\nu}} = \\ = \frac{1}{\lambda+\mu^\nu}\end{aligned} \tag{5.137}$$

We are now able to *indirectly* calculate the Laplace transform from the renewal representation of the process, as:

$$\int_0^\infty e^{-\mu t} Pr\left\{N_\nu(t)=k\right\}dt = \frac{\mu^{\nu-1}\lambda^k}{(\lambda+\mu^\nu)^{k+1}} \tag{5.138}$$

owing to the renewal representation

$$\begin{aligned} Pr\left\{N_\nu(t)=k\right\} = \\ = Pr\left\{U_1+\ldots+U_k<t, U_1+\ldots+U_{k+1}>t\right\} = \\ = Pr\left\{U_1+\ldots+U_k<t\right\} - Pr\left\{U_1+\ldots+U_{k+1}<t\right\} = \\ = Pr\{T_k^\nu<t\} - \Pr\{T_{k+1}^\nu<t\}\end{aligned} \tag{5.139}$$

because $\Pr\{A\cap B\} = \Pr\{A\} - \Pr\{A\cap B^c\}$.

Indeed,

$$
\begin{aligned}
&\int_0^\infty e^{-\mu t}\Pr\{N_\nu(t)=k\}dt = \\
&= \int_0^\infty e^{-\mu t}[\Pr\{T_k^\nu < t\} - \Pr\{T_{k+1}^\nu < t\}]dt = \\
&= \int_0^\infty e^{-\mu t}\left[\Pr\{U_1+\ldots+U_k<t\}+\right. \\
&\left.-\Pr\{U_1+\ldots+U_k+U_{k+1}<t\}\right]dt = \\
&= \int_0^\infty e^{-\mu t}\int_0^t \Pr\{U_1+\ldots+U_k\in ds\}dt+ \\
&-\int_0^\infty e^{-\mu t}\int_0^t \Pr\{U_1+\ldots+U_k+U_{k+1}\in ds\}dt = \\
&= \int_0^\infty \Pr\{U_1+\ldots+U_k\in ds\}\int_s^\infty e^{-\mu t}dt+ \\
&-\int_0^\infty \Pr\{U_1+\ldots+U_k+U_{k+1}\in ds\}\int_s^\infty e^{-\mu t}dt = \\
&= \frac{1}{\mu}\int_0^\infty e^{-\mu s}\Pr\{U_1+\ldots+U_k\in ds\}+ \\
&-\frac{1}{\mu}\int_0^\infty e^{-\mu s}\Pr\{U_1+\ldots+U_k+U_{k+1}\in ds\} = \\
&= \frac{1}{\mu}\left[\int_0^\infty e^{-\mu s}\Pr\{U_1\in ds\}\right]^k + \\
&-\frac{1}{\mu}\left[\int_0^\infty e^{-\mu s}\Pr\{U_1\in ds\}\right]^{k+1} = \\
&= \frac{1}{\mu}\left[\left(\frac{\lambda}{\lambda+\mu^\nu}\right)^k - \left(\frac{\lambda}{\lambda+\mu^\nu}\right)^{k+1}\right] = \\
&= \left(\frac{\lambda}{\lambda+\mu^\nu}\right)^k \frac{1}{\mu}\left(1-\frac{\lambda}{\lambda+\mu^\nu}\right) = \\
&= \left(\frac{\lambda}{\lambda+\mu^\nu}\right)^k \frac{\mu^{\nu-1}}{\lambda+\mu^\nu} = \frac{\mu^{\nu-1}\lambda^k}{(\lambda+\mu^\nu)^{k+1}}
\end{aligned}
\tag{5.140}
$$

We now show that it is equal to the Laplace transform of $\Pr\{N_\nu(t)=k\}$. In fact, the direct computation of the Laplace transform reads as follows:

$$\begin{aligned}
\int_0^\infty \Pr\{N_\nu(t)=k\}dt &= \\
= \int_0^\infty e^{-\mu t}\sum_{h=k}^\infty (-1)^{h-k}\binom{h}{k}\frac{(\lambda t^\nu)^h}{\Gamma(\nu h+1)} &= \\
= \sum_{h=k}^\infty (-1)^{h-k}\binom{h}{k}\frac{1}{\Gamma(\nu h+1)}\cdot\int_0^\infty e^{-\mu t}(\lambda t^\nu)dt &= \\
= \sum_{h=k}^\infty (-1)^{h-k}\binom{h}{k}\frac{\lambda^h}{\Gamma(\nu h+1)}\frac{\Gamma(\nu h+1)}{\mu^{\nu h+1}} &= \\
= \frac{1}{\mu}\sum_{h=k}^\infty (-1)^{h-k}\binom{h}{k}\left(\frac{\lambda}{\mu^\nu}\right)^h &= \\
(h-k=h') &= \\
= \frac{1}{\mu}\sum_{h'=0}^\infty (-1)^{h'}\binom{h'+k}{k}\left(\frac{\lambda}{\mu^\nu}\right)^{k+h'} &= \\
= \frac{1}{\mu}\left(\frac{\lambda}{\mu^\nu}\right)^k\sum_{h'=0}^\infty \binom{h+k}{h}\left(-\frac{\lambda}{\mu^\nu}\right)^h &= \\
= \frac{1}{\mu}\left(\frac{\lambda}{\mu^\nu}\right)^k\sum_{h=0}^\infty (-1)^h\binom{-(k+1)}{h}\left(-\frac{\lambda}{\mu^\nu}\right)^h &= \\
= \frac{1}{\mu}\left(\frac{\lambda}{\mu^\nu}\right)^k\left(1+\frac{\lambda}{\mu^\nu}\right)^{-(k+1)} &= \frac{\mu^{\nu-1}\lambda^k}{(\lambda+\mu^\nu)^{k+1}}
\end{aligned} \tag{5.141}$$

because $\left(1+\frac{\lambda}{\mu^\nu}\right)^{-(k+1)} = \sum_{h=0}^\infty \binom{-(k+1)}{h}\left(-\frac{\lambda}{\mu^\nu}\right)^h$ and $\binom{-(k+1)}{h} = (-1)^h\frac{(h+k)!}{k!h!}$.

The time-fractional Poisson process can also be represented as a Poisson process with a time change obtained by the "inverse" of a stable subordinator (see Meerschaert et al, 2011). More precisely, it admits the following representation:

$$N_\nu(t) \stackrel{D}{=} N(L^\nu(t)), \qquad 0<\nu\le 1, \qquad t\ge 0 \tag{5.142}$$

where N is a homogeneous Poisson process, and $L^{\nu}(t)$ is the inverse of a stable subordinator $S^{\nu}(t)$, i.e.

$$Pr\{L^{\nu}(t) > s\} = Pr\{S^{\nu}(s) < t\} \tag{5.143}$$

For the hitting time of the time-fractional Poisson process

$$\tau_k^{(\nu)} = \inf(t > 0 : N_{\nu}(t) = k), \quad k \geq 1, \nu \in (0, 1] \tag{5.144}$$

Beghin and Orsingher (2010) provide the expression

$$\begin{aligned} \Pr\{\tau_k^{(\nu)} \in ds\} &= \lambda \sum_{h=0}^{\infty} \binom{-k}{h} \lambda^h \frac{s^{\nu(k+h)-1}}{\Gamma(\nu(k+h))} ds \\ &= \lambda^k s^{\nu k-1} E_{\nu,\nu k}^{k}(-\lambda s^{\nu}) ds \end{aligned} \tag{5.145}$$

where $E_{\alpha,\beta}^{\gamma}(x) = \sum_{h=0}^{\infty} \frac{(\gamma)_h x^h}{h!\Gamma(\alpha h+\beta)}$ is a generalized Mittag-Leffler function (and $(\gamma)_h$ is the Pochhammer symbol).

Remark 5.5.

The time-fractional Skellam process

The time-fractional Skellam process is expressed in terms of the difference of two time-fractional Poisson processes (see Buchak and Sakhno, 2019)

$$f_{Sk}(t) = N_1(L^{(1)}(t)) - N_2(L^{(2)}(t))$$

where $L^{(1)}(t)$ and $L^{(2)}(t)$ are two independent inverse stable subordinators of indexes ν_1 and ν_2 both in $(0, 1)$.

5.4.1 Estimation of the Time-fractional Poisson Process

The section presents point estimation of the process as developed after the 2009 article of Beghin and Orsingher, with an application to earthquake data in the context of ruin theory.

5.4.1.1 Estimators and Confidence Intervals

Cahoy et al. (2010) developed a method-of-moments estimator for the parameters (λ, ν) of the time-fractional Poisson process.

The estimation exploits the renewal representation of the process by conducting inference on the (logarithm of) inter-arrival time $\ln T_k^{(\nu)}$, i.e. (the logarithm of) the time between the k-1*th* and k-*th* events. Its mean and variance are:

$$\begin{cases} \mathbb{E}[\ln T_k(\nu)] = -\left(\frac{\ln \lambda}{\nu} + \mathbb{C}\right) \\ \mathbb{V}ar[\ln T_k(\nu)] = \left(\frac{\pi}{\sqrt{3}\nu}\right)^2 - \frac{\pi^2}{6} \end{cases} \tag{5.146}$$

Inverting the above equations, the estimators are obtained as

$$\begin{cases} \hat{\lambda} = \exp\left(-\hat{\nu}\left(\hat{\mu}_{\ln\{T_k^{(\nu)}\}} + \mathbb{C}\right)\right) \\ \hat{\nu} = \frac{\pi}{\sqrt{3\left(\hat{\sigma}^2_{\ln T_k^{(\nu)}} + \frac{\pi^2}{6}\right)}} \end{cases} \tag{5.147}$$

where $\mu_{\ln\{T_k^{(\nu)}\}}$ and $\sigma^2_{\ln\{T_k^{(\nu)}\}}$ represent the expected value and variance of the r.v. $\ln\{T_k^{(\nu)}\}$ and $\mathbb{C}$ is Euler-Mascheroni constant, i.e. $\mathbb{C} \approx 0.57721$[4].

Remark 5.6.
Note that Cahoy (2013) further expands their method of moments estimator with explicit reference to the *Mittag-Leffler distribution*, which is the inter-time law for the time-fractional Poisson process.

Confidence interval. Cahoy et al. (2010) also provide the asymptotic distributions of the parameter estimates $\hat{\lambda}$ and $\hat{\nu}$, w.r.t. the sample size n, which are Normal and unbiased. In particular, for the fractionality parameter ν, the asymptotics is:

$$\sqrt{n}(\hat{\nu} - \nu) \sim N\left(0, \frac{\nu^2(32 - 20\nu^2 - \nu^4)}{40}\right) \tag{5.148}$$

The resulting $(1-\alpha)100\%$ confidence interval, for the estimate $\hat{\nu}$, is

$$\hat{\nu} \pm z_{\alpha/2}\sqrt{\frac{\hat{\nu}^2(32 - 20\hat{\nu}^2 - \hat{\nu}^4)}{40n}} \tag{5.149}$$

[4]The Euler-Mascheroni constant $\mathbb{C}$ is related to the partial sum of the harmonic series, i.e. $\sum_{h=1}^{n} \frac{1}{h} = \ln n + \mathbb{C} + o(1)$.

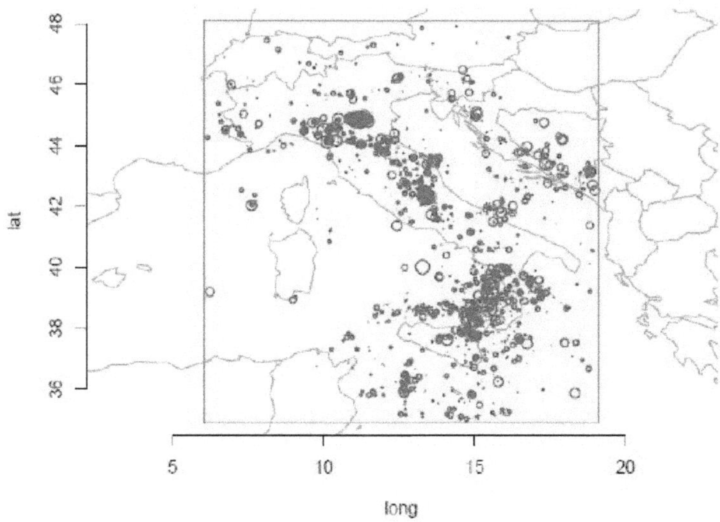

FIGURE 5.8
Epicenters of 2158 earthquake shocks with magnitude momentum (Mw) > 3.0 (i.e., recorded by seismographs and felt by the population), 2005–2013.

where $\Pr\{N(0,1) > z_{\alpha/2}\} = \Pr\{N(0,1) < -z_{\alpha/2}\} = \frac{\alpha}{2}$ (e.g. $z_{0.05/2} = 1.96$).

5.4.1.2 An Application to Earthquakes and Ruin Theory

Estimation on earthquake data. Estimation to the Italian earthquake catalog[5] (earthquakes from 1600) results in a "*fractionality parameter*" less than unity, confirming data clustering, in line with the nature of *aftershocks*.

An application to ruin probability. Biard and Saussereau (2014) and Constantinescu et al. (2019) apply the time-fractional Poisson process to the ruin probability $\psi(u)$ in the context of the Collective Risk Model (CRM). A CRM models the accumulation

[5]For general data on earthquakes, it is possible to exploit the R package *ETAS*: for example, Fig. 5.8 shows data on Italian seismicity.

TABLE 5.1
Estimation on Italian earthquakes.[6]

parameter estimates	$\hat{\nu}$	$\hat{\lambda}$
MoM-1 (Cahoy et al., 2010)	0.49	0.66
MoM-2 (Cahoy et al., 2013)	0.85	0.43

and decumulation of (risky) capital in an insurance fund by means of a reserve (of capital) process $R(t) = u + \pi t - \sum_{i=1}^{N(t)} X_i$, where $u = R(0) > 0$ is the initial capital endowment, π is the premium intensity (such that πt is a monetary amount representing premium inflows by time t), and the X_i's are i.i.d. insurance claims filed within time t, occurring at the pace of a Poisson process $N(t)$.

In the work of Biard and Constantinescu, the arrival process is generalized as a (time) fractional Poisson process $N_\nu(t)$. More precisely, they provide *analytic* formulas of the *infinite-time* ruin probability (as a function of initial capital u) $\psi_\nu(u) = \Pr\{R_t < 0\}$, $t \geq 0$, in the extended case of a *heavy-tailed*[7] number $N_\nu(t)$ of *light-tailed* risks X (in particular, with finite mean μ_X).

More precisely, Biard and Saussereau (2014) derive the *asymptotic* ruin probability $\psi_\nu(u)$ as

$$\psi_\nu(u) = \left(1 - \frac{x^*}{\mu_X}\right) e^{-x^* u} \qquad \nu < 1 \tag{5.150}$$

where x^* solves (unique solution) the equation $x - \mu_X + \lambda x^{1-\nu}\pi^{-\nu} = 0$.

Constantinescu et al. (2019) obtain – by means of fractional calculus operators – the *asymptotic* ruin probability as in the equation (5.150) in Biard and Saussereau (2014) valid for an exponential claim sizes. The asymptotic ruin probability is monotone increasing in both the λ and ν parameters.

[6]Source: https://emidius.mi.ingv.it

[7]Note that the process inter-times $\Delta T_k^{(\nu)}$ are long-tailes, i.e. $\Pr\{\Delta T_k^{(\nu)} > t\} \sim Ct^{-\nu}$, $C > 0$; more precisely, $\Pr\{\Delta T_k^{(\nu)} > t\} = E(-\lambda t^\nu)$, where $E(\cdot)$ is the Mittag-Leffler function.

For comparison, in the simple Compound Poisson model with: aggregate claims $\sum_{i=1}^{N(t)} X_i$, Poisson arrival process of rate $\lambda > 0$ and Exponential claims of rate $\mu > 0$, the asymptotic ruin probability $\psi(u)$ is of exponential type as well and equals (see Albrecher and Asmussen, 2010) $\psi_(u) = \frac{\lambda}{\mu} e^{-(\mu-\lambda)u}$ (unit premium intensity $\pi = 1)^8$.

Remark 5.7.
Note that Biard and Saussereau (2014) also derives an analytic expression for the ruin probability in *finite-time* under light-tailed claims distributions.

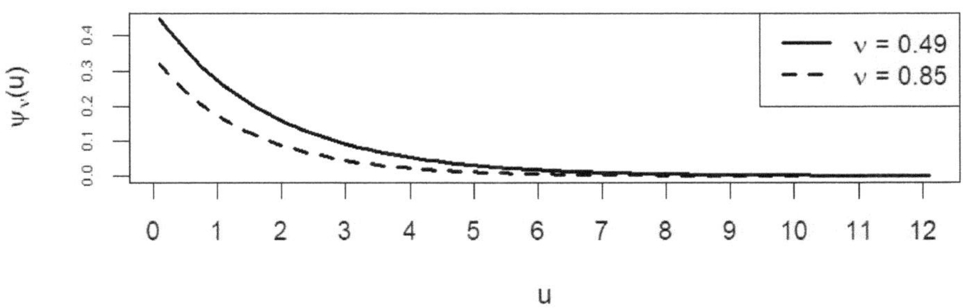

FIGURE 5.9
Asymptotic ruin probability $\psi_\nu(u)$ for ($\nu = 0.49$, $\lambda = 0.66$) (above) and ($\nu = 0.85$, $\lambda = 0.43$) (below). Note that the effect of λ prevails over ν and the ruin probability with $\lambda = 0.66$ (and $\nu = 0.49$) is over the one with $\lambda = 0.43$ (and $\nu = 0.85$). The premium intensity $\pi = 1.2$).

For $\nu < 1$ the ruin probability is lower (intuitively, it corresponds to infinite expected waiting times) than for $\nu = 1$.

[8]Note that equation (5.150) for $\nu = 1$ greatly simplifies, and in particular turns out to be an explicit function of the Compound Poisson claims model parameters λ and μ.

5.5 The Space-time Fractional Poisson Process

Introduction. We introduce the space-fractional Poisson process by starting from the equation governing the state probabilities

$$\frac{d^\nu p_k^{\alpha,\nu}(t)}{dt^\nu} = -\lambda^\alpha (I-B)^\alpha p_k^{\alpha,\nu}(t) \tag{5.151}$$

with $0<\nu<1$, $0<\alpha<1$, where I is the identity operator, and B is the shift operator.

The distribution $p_k^{\alpha,\nu}(t)$, $k \geq 0$, of the space-fractional Poisson process is obtained from (5.151) with initial conditions

$$p_k^{\alpha,\nu}(0) = \begin{cases} 1 & k=0 \\ 0 & k \geq 1 \end{cases} \tag{5.152}$$

The probability generating function of the space-time fractional Poisson process can be constructed starting from (5.151), written in its explicit form, as

$$\frac{d^\nu}{dt^\nu} p_k^{\alpha,\nu}(t) = -\lambda^\alpha \sum_{h=0}^{k} \frac{\Gamma(\alpha+1)}{h!\Gamma(\alpha+1-h)}(-1)^h p_{k-h}^{\alpha,\nu}(t); \tag{5.153}$$

in fact

$$\begin{aligned}\frac{d^\nu G_{\alpha,\nu}(u,t)}{dt^\nu} &= \sum_{k=0}^{\infty} u^k \frac{d^\nu}{dt^\nu} p_k^{\alpha,\nu}(t) = \\ = -\lambda^\alpha \sum_{k=0}^{\infty} u^k \sum_{h=0}^{k} &\frac{\Gamma(\alpha+1)}{h!\Gamma(\alpha+1-h)}(-1)^h p_{k-h}^{\alpha,\nu}(t) = \\ = -\lambda^\alpha \sum_{h=0}^{\infty}\sum_{k=h}^{\infty} &\frac{u^k\Gamma(\alpha+1)}{h!\Gamma(\alpha+1-h)}(-1)^h p_{k-h}^{\alpha,\nu}(t) = \\ &(k-h=k') \\ = -\lambda^\alpha \sum_{h=0}^{\infty}\sum_{k'=0}^{\infty} &\frac{u^{h+k'}\Gamma(\alpha+1)}{\Gamma(\alpha+1-h)}(-1)^h p_{k'}^{\alpha,\nu}(t) = \\ = -\lambda^\alpha \sum_{h=0}^{\infty} &\frac{(-u)^h\Gamma(\alpha+1)}{h!\Gamma(\alpha+1-h)} \cdot \sum_{k=0}^{\infty} u^k p_k^{\alpha,\nu}(t) = \\ &= -\lambda^\alpha G_{\alpha,\nu}(u,t)(1-u)^\alpha\end{aligned} \tag{5.154}$$

Thus, the probability generating function of the space-time fractional Poisson process satisfies

$$\frac{\partial^\nu G_{\alpha,\nu}(u,t)}{\partial t^\nu} = -\lambda^\alpha (1-u)^\alpha G_{\alpha,\nu}(u,t) \tag{5.155}$$

subject to the initial condition $G_{\alpha,\nu}(u,0) = 1$.

Thus[9]

$$\begin{aligned} G_{\alpha,\nu}(u,t) &= E_{\nu,1}(-\lambda^\alpha (1-u)^\alpha t^\nu) = \\ &= \sum_{h=0}^{\infty} \frac{(-\lambda^\alpha (1-u)^\alpha t^\nu)^h}{\Gamma(\nu h+1)} = \\ &= \sum_{h=0}^{\infty} \frac{(-\lambda^\alpha t^\nu)^h}{\Gamma(\nu h+1)} (1-u)^{\alpha h} = \\ &= \sum_{h=0}^{\infty} \frac{(-\lambda^\alpha t^\nu)^h}{\Gamma(\nu h+1)} \cdot \sum_{k=0}^{\infty} \frac{(-u)^k \Gamma(\alpha h+1)}{k!\Gamma(\alpha h+1-k)} = \\ &= \sum_{k=0}^{\infty} \frac{u^k}{k!}(-1)^k \cdot \sum_{h=0}^{\infty} \frac{(-\lambda^\alpha t^\nu)^h}{\Gamma(\nu h+1)} \frac{\Gamma(\alpha h+1)}{\Gamma(\alpha h+1-k)} \end{aligned} \tag{5.156}$$

The probability $p_k^{\alpha,\nu}(t)$ is therefore

$$p_k^{\alpha,\nu}(t) = \frac{(-1)^k}{k!} \sum_{h=0}^{\infty} \frac{(-\lambda^\alpha t^\nu)^h \Gamma(\alpha h+1)}{\Gamma(\nu h+1)\Gamma(\alpha h+1-k)} \tag{5.157}$$

A space-time fractional Poisson process N_ν^α, $\alpha, \nu \in (0,1)$, can be easily understood by means of the subordination of an ordinary Poisson process to both a Stable and *then* its inverse stable subordinator. As such, it can be represented as $N_\nu^\alpha(t) = N(S_\alpha(L_\alpha(t)))$, $t \geq 0$, where $D_\alpha(t)$ is an independent α-Stable subordinator and $L_\alpha(t)$ is its inverse.

We refer to Orsingher and Polito (2012b), where the process is defined as an extension to the space-fractional Poisson (see also Orsingher and Polito, 2013).

[9]Since the Mittag-Leffler function is the auto-function of the Dzherbashyan-Caputo fractional differential operator.

Quite naturally, on one side, the "space-fractional" dimension of the process involves the operator $(1-B)^{\alpha}$ for its construction, as it appears in the (fractional) Cauchy problem for the state probabilities $p_k^{\alpha,\nu}(t)$, $k \geq 0$, i.e.

$$\frac{d^{\nu} p_k^{\alpha,\nu}(t)}{dt^{\nu}} = -\lambda^{\alpha}(1-B)^{\alpha} p_k^{\alpha,\nu}(t) \tag{5.158}$$

where, on the other side, a (Caputo) time-fractional derivative operator takes into accounts the time-fractional dimension, subject to the initial condition

$$p_k^{\alpha,\nu}(0) = \begin{cases} 0 & k \geq 1 \\ 1 & k = 0 \end{cases} \tag{5.159}$$

The "time-fractional" and "space-fractional" dimensions appear more evident in state probabilities $p_k^{\alpha,\nu}(t) = \Pr\{N_{\nu}^{\alpha}(t) = k\}$, which are given as

$$p_k^{\alpha,\nu}(t) = \frac{(-1)^k}{k!} \sum_{h=0}^{\infty} \frac{(-\lambda^{\alpha} t^{\nu})^h}{\Gamma(\nu h + 1)} \frac{\Gamma(\alpha h + 1)}{\Gamma(\alpha h + 1 - k)}, \qquad k \geq 1, \quad \alpha \in (0,1], \quad \nu \in (0,1] \tag{5.160}$$

(see Orsingher and Polito, 2012b) where the dependence on the time derivative is expressed bu ν and the dependence on the space operator is given by α. Indeed, setting $\alpha = 1$ and $\nu = 1$, the time-fractional and space-fractional Poisson processes are recovered, respectively. See also Buchak and Sakhno (2017).

The probability generating function $G_{\alpha,\nu}(u)$ obtains as solution of

$$\begin{cases} \frac{d^{\nu}}{dt^{\nu}} G_{\alpha,\nu}(u,t) = -\lambda^{\alpha}(1-u)^{\alpha} G_{\alpha,\nu}(u,t) \\ G_{\alpha,\nu}(u,0) = 1 \end{cases} \tag{5.161}$$

that is

$$G_{\alpha,\nu}(u) = E_{\nu,1}(-\lambda^{\alpha}(1-u)^{\alpha} t^{\nu}) \qquad |u| \leq 1 \tag{5.162}$$

where the $E_{\nu,1}(x) = E_{\nu}(x)$ is indeed a one-parameter Mittag-Leffler function.

The p.g.f. $G_{\alpha,\nu}(u)$ has a probabilistic interpretation with respect to the event $(\min_{1\leq k\leq N_\nu(t)} U_k^{\frac{1}{\alpha}} \geq 1-u)$, where the random variables U_k, $k\geq 1$, are i.i.d uniforms, and $N_\nu(t)$ is the time-fractional Poisson process, i.e.

$$G_{\alpha,\nu}(u) = \Pr\left\{\min_{1\leq k\leq N_\nu(t)} U_k^{\frac{1}{\alpha}} \geq 1-u\right\} \qquad |u|<1. \tag{5.163}$$

Subordination. The space-time fractional Poisson process has the following (time-changed) representation

$$N^{\alpha,\nu}(t) = N(S^\alpha(L^\nu(t))) \tag{5.164}$$

where $N(t)$ is a homogeneous Poisson process, $S^\alpha(t)$ is a stable subordinator and L^ν is the inverse of a stable subordinator, independent of $S^\alpha(t)$.

In order to prove (5.164), we consider the probability generating function $G_{\alpha,\nu}(u,t)$, i.e.

$$\int_0^\infty e^{-\mu t}\mathbb{E}\left[u^{N(S_\alpha(L_\nu(t)))}\right] dt \tag{5.165}$$

But before that, we must derive the Laplace transform of the inverse stable subordinator.

The Laplace transform of the inverse subordinator. Since

$$\Pr\{L_\nu(t)>s\} = \Pr\{S_\nu(s)<t\} = \int_0^t \Pr\{S_\nu(s)\in dw\} \tag{5.166}$$

then

$$\Pr\{L_\nu(t)\in ds\} = -\frac{d}{ds}\left\{\int_0^t \Pr\{S_\nu(s)\in dw\}\right\} \tag{5.167}$$

and we have that

$$\begin{aligned}
&\int_0^\infty e^{-\mu t}\Pr\{L_\nu(t)\in ds\}dt = \\
&= \int_0^\infty e^{-\mu t}\left\{-\frac{d}{ds}\int_0^t \Pr\{S_\nu(s)\in dw\}\right\}dt = \\
&= -\frac{d}{ds}\int_0^\infty e^{-\mu t}dt\cdot\int_0^t \Pr\{S_\nu(s)\in dw\} = \\
&= -\frac{d}{ds}\int_0^\infty \Pr\{S_\nu(s)\in dw\}\cdot\int_w^\infty e^{-\mu t}dt = \\
&= -\frac{d}{ds}\int_0^\infty \frac{e^{-\mu w}}{\mu}\Pr\{S_\nu(s)\in dw\} = \\
&= -\frac{d}{ds}\frac{e^{-\mu^\nu s}}{\mu} = \mu^{\nu-1}e^{-\mu^\nu s}
\end{aligned} \tag{5.168}$$

Nested time-changed representation. We can now prove (5.164), by considering the Laplace transform of the probability generating function

$$\begin{aligned}
&\int_0^\infty e^{-\mu t}\mathbb{E}\left[u^{N(S_\alpha(L_\nu(t)))}\right]dt = \\
&= \int_0^\infty e^{-\mu t}\mathbb{E}\left[u^{N(S_\alpha(L_\nu(t)))}\mid S_\alpha(L_\nu(t))\right]dt = \\
&= \int_0^\infty e^{-\mu t}\mathbb{E}\left[e^{-\lambda S_\alpha(L_\nu(t))(1-u)}\right]dt = \\
&= \int_0^\infty e^{-\mu t}\mathbb{E}\left[\mathbb{E}\left[e^{-\lambda S_\alpha(L_\nu(t))(1-u)}\mid L_\nu(t)\right]\right]dt = \\
&= \int_0^\infty e^{-\mu t}\mathbb{E}\left[e^{-\lambda^\alpha(1-u)^\alpha L_\nu(t)}\right]dt
\end{aligned} \tag{5.169}$$

The inner expected value reads $\int_0^\infty e^{-\lambda^\alpha(1-u)^\alpha s}\Pr\{L_\nu(t)\in ds\}$, thus

$$\begin{aligned}
&\int_0^\infty e^{-\mu t}\int_0^\infty e^{-\lambda^\alpha(1-u)^\alpha s}\Pr\{L_\nu(t)\in ds\}dt = \\
&= \int_0^\infty e^{-\lambda^\alpha(1-u)^\alpha s}\int_0^\infty e^{-\mu t}\Pr\{L_\nu(t)\in ds\}dt = \\
&= \int_0^\infty e^{-\lambda^\alpha(1-u)^\alpha s}\cdot\mu^{\nu-1}e^{-\mu^\nu s}ds = \frac{\mu^{\nu-1}}{\mu^\nu+\lambda^\alpha(1-u)^\alpha}
\end{aligned} \tag{5.170}$$

where we exploit the result in (5.168) for the Laplace transform of the inverse of a stable subordinator.

Since we must prove that

$$N^{\alpha,\nu}(t) = N(S_\alpha(L_\nu(t))) \tag{5.171}$$

we need to show that

$$\int_0^\infty e^{-\mu t}\mathbb{E}\left[u^{N^{\alpha,\nu}(t)}\right] dt = \int_0^\infty e^{-\mu t}\mathbb{E}\left[u^{N(S_\alpha(L_\nu(t)))}\right] dt \tag{5.172}$$

We have just shown that the RHS

$$\int_0^\infty e^{-\mu t}\mathbb{E}\left[u^{N(S_\alpha(L_\nu(t)))}\right] dt = \frac{\mu^{\nu-1}}{\mu^\nu + \lambda^\alpha(1-u)^\alpha} \tag{5.173}$$

It remains to evaluate the LHS, i.e.

$$\begin{aligned} &\int_0^\infty e^{-\mu t}\mathbb{E}\left[u^{N^{\alpha,\nu}(t)}\right] dt = \\ &= \int_0^\infty e^{-\mu t} E_{\nu,1}(-\lambda^\alpha(1-u)^\alpha t^\nu)dt = \\ &= \sum_{k=0}^\infty \frac{(-\lambda^\alpha(1-u)^\alpha)^k}{\Gamma(\nu k+1)} \int_0^\infty e^{-\mu t}t^{\nu k}dt = \\ &= \sum_{k=0}^\infty \frac{(-\lambda^\alpha(1-u)^\alpha)^k}{\Gamma(\nu k+1)} \cdot \frac{\Gamma(\nu k+1)}{\mu^{\nu k+1}} = \\ &= \frac{1}{\mu}\left[\frac{1}{1+\frac{\lambda^\alpha(1-u)^\alpha}{\mu^\nu}}\right] = \frac{\mu^{\nu-1}}{\mu^\nu + \lambda^\alpha(1-u)^\alpha} \end{aligned} \tag{5.174}$$

This concludes the check of (5.164).

5.6 Fractional Birth-and-death Processes

We provide the main probabilistic features for the fractional generalizations studied by Orsingher and coauthors.

5.6.1 The Fractional Pure-birth Process

The non-linear process. The non-linear fractional pure-birth process has been studied by Orsingher and Polito (2010) as a generalization of the pure-birth process, by considering the fractional derivative w.r.t. time in the Cauchy problem for the state probabilities, i.e.:

$$\frac{d^\nu p_k^\nu}{dt^\nu} = -\lambda_k p_k^\nu + \lambda_{k-1} p_{k-1}^\nu, \quad k \geq 1, \quad 0 < \nu < 1 \tag{5.175}$$

subject to the initial conditions $p_1^\nu(0) = 1$ and $p_k^\nu(0) = 0$ for $k \geq 2$.

Its solution is given in the following theorem (see Orsingher and Polito (2010), Theorem 2.1):

Theorem 5.8. *The state probabilities $p_k^\nu(t) = Pr\{\mathcal{N}_\nu(t) = k\}$ of the non-linear birth process $\mathcal{N}_\nu(t)$ read*

$$p_k^\nu(t) = \begin{cases} E_{\nu,1}(-\lambda_1 t^\nu) & k = 1 \\ \prod_{j=1}^{k=1} \lambda_j \sum_{m=1}^{k} \left\{ \frac{1}{\prod_{h=1,h\neq m}^{k} (\lambda_h - \lambda_m)} \right\} E_{\nu,1}(-\lambda_m t^\nu), & k \geq 2 \end{cases} \tag{5.176}$$

where $E_{\nu,1}(x) = \sum_{k=0}^{\infty} \frac{x^k}{\Gamma(\nu k+1)}$ ($x \in \mathbb{R}$ and $0 < \nu < 1$) is the two-parameter Mittag-Leffler function.

Proof. For $k = 1$, applying the Laplace transform $\mathcal{L}_k(\mu) \equiv \mathcal{L}[p_k(t)](\mu) = \int_0^\infty e^{-\mu t} p_k(t) dt$ to the fractional differential equation

$$\frac{d^\nu}{dt^\nu} p_1^\nu(t) = -\lambda_1 p_1^\nu(t) \tag{5.177}$$

one obtains (with $p_1(0) = 1$)

$$\begin{cases} \mu^\nu \mathcal{L}_\infty(\mu) - \mu^{\nu-1} = -\lambda_1 \mathcal{L}_1(\mu) \\ \mathcal{L}_1(\mu) = \frac{\mu^{\nu-1}}{\mu^\nu + \lambda_1} \end{cases} \tag{5.178}$$

and, hence, applying the inverse Laplace transform, we have that the solution is

$$p_1^\nu(t) = E_{\nu,1}(-\lambda_1 t^\nu) \tag{5.179}$$

For $k = 2$, the substitution of the solution $p_1^\nu(t)$ into (5.175) yields

$$\begin{cases} \frac{d^\nu p_2^\nu(t)}{dt^\nu} = -\lambda_2 p_2^\nu(t) + \lambda_1 E_{\nu,1}(-\lambda_1 t^\nu) \\ p_2^\nu(0) = 0 \end{cases} \tag{5.180}$$

Applying the Laplace transform to both sides (remembering that $p_2(0) = 0$) yields

$$\mu^\nu \mathcal{L}_2 = -\lambda_2 \mathcal{L}_2 + \lambda_1 \mathcal{L}\left[E_{\nu,1}(-\lambda_1 t^\nu)\right] \tag{5.181}$$

since (see Appendix D) the Laplace transform of the last Mittag-Leffler function reads

$$\mathcal{L}\left[E_{\nu,1}(-\lambda_1 t^\nu)\right](\mu) = \frac{\mu^{\nu-1}}{\lambda_1 + \mu^\nu} \tag{5.182}$$

Then

$$\begin{aligned} \mathcal{L}_2(\mu) &= \\ &= \frac{\lambda_1}{\lambda_2 + \mu^\nu} \frac{\mu^{\nu-1}}{\lambda_1 + \mu^\nu} = \\ &= \frac{\lambda_1}{\lambda_2 - \lambda_1} \left[\frac{1}{\lambda_1 + \mu^\nu} - \frac{1}{\lambda_2 + \mu^\nu}\right] \mu^{\nu-1} \end{aligned} \tag{5.183}$$

and inverting the Laplace transform yields the state probability $p_2^\nu(t)$

$$p_2^\nu(t) = \frac{\lambda_1}{\lambda_2 - \lambda_1} \left[E_{\nu,1}(-\lambda_1 t^\nu) - E_{\nu,1}(-\lambda_2 t^\nu)\right] \tag{5.184}$$

For $k = 3$ equation (5.175) now reads

$$\frac{d^\nu}{dt^\nu} p_3^\nu(t) = -\lambda_3 p_3^\nu(t) + \lambda_2 \left(\frac{\lambda_1}{\lambda_2 - \lambda_1} \left[E_{\nu,1}(-\lambda_1 t^\nu) - E_{\nu,1}(-\lambda_2 t^\nu)\right]\right) \tag{5.185}$$

subject to the initial condition $p_3^\nu(0) = 0$.

The same kind of computations lead to

$$\begin{aligned}\mathcal{L}_3(\mu) = &\\ &\lambda_1\lambda_2\mu^{\nu-1}\left[\frac{1}{(\lambda_2-\lambda_1)(\lambda_3-\lambda_1)}\frac{1}{\lambda_1+\mu^\nu}+\right.\\ &+\frac{1}{(\lambda_1-\lambda_2)(\lambda_3-\lambda_2)}\frac{1}{\lambda_2+\mu^\nu}\\ &\left.+\frac{1}{(\lambda_1-\lambda_3)(\lambda_2-\lambda_3)}\frac{1}{\lambda_3+\mu^\nu}\right]\end{aligned} \tag{5.186}$$

from which it is clear that

$$\begin{aligned}p_3(t) = &\\ &\lambda_1\lambda_2\cdot\left[\frac{1}{(\lambda_2-\lambda_1)(\lambda_3-\lambda_1)}E_{\nu,1}(-\lambda_1 t^\nu)+\right.\\ &+\frac{1}{(\lambda_1-\lambda_2)(\lambda_3-\lambda_2)}E_{\nu,1}(-\lambda_2 t^\nu)+\\ &\left.+\frac{1}{(\lambda_1-\lambda_3)(\lambda_2-\lambda_3)}E_{\nu,1}(-\lambda_3 t^\nu)\right]\end{aligned} \tag{5.187}$$

The general form begins now to arise. We now hint at the technique involved for the case $k=4$, but refer to Orsingher and Polito (2010) for further details about the case $k=4$, which is instructive for the cases $k \geq 4$.

The Laplace transform of the equation

$$\begin{aligned}\frac{d^\nu}{dt^\nu}p_4^\nu(t) = &\\ &-\lambda_4 p_4^\nu(t)+\\ &+\lambda_1\lambda_2\lambda_3\cdot\left[\frac{1}{(\lambda_2-\lambda_1)(\lambda_3-\lambda_1)}E_{\nu,1}(-\lambda_1 t^\nu)+\right.\\ &+\frac{1}{(\lambda_1-\lambda_2)(\lambda_3-\lambda_2)}E_{\nu,1}(-\lambda_2 t^\nu)+\\ &\left.+\frac{1}{(\lambda_1-\lambda_3)(\lambda_2-\lambda_3)}E_{\nu,1}(-\lambda_3 t^\nu)\right]\end{aligned} \tag{5.188}$$

subject to the initial condition $p_4^\nu(0) = 0$, becomes

$$\begin{aligned}
\mathcal{L}_4(\mu) = \\
= \lambda_1\lambda_2\lambda_3\mu^{\nu-1}\times \\
\times\left[\frac{1}{(\lambda_2-\lambda_1)(\lambda_3-\lambda_1)(\lambda_4-\lambda_1)}\left(\frac{1}{\lambda_1+\mu^\nu}-\frac{1}{\lambda_4+\mu^\nu}\right)+\right. \\
+\frac{1}{(\lambda_1-\lambda_2)(\lambda_3-\lambda_2)(\lambda_4-\lambda_2)}\left(\frac{1}{\lambda_2+\mu^\nu}-\frac{1}{\lambda_4+\mu^\nu}\right)+ \\
\left.+\frac{1}{(\lambda_1-\lambda_3)(\lambda_2-\lambda_3)(\lambda_4-\lambda_3)}\left(\frac{1}{\lambda_3+\mu^\nu}-\frac{1}{\lambda_4+\mu^\nu}\right)\right].
\end{aligned} \tag{5.189}$$

After some manipulations of the involved determinants[10], the Laplace transform simplifies into a structure where the sum of Laplace transforms of Mittag-Leffler functions arises, i.e.

$$\begin{aligned}
\mathcal{L}_4(\mu) = \\
= \lambda_1\lambda_2\lambda_3\mu^{\nu-1}\cdot\left[\frac{1}{(\lambda_2-\lambda_1)(\lambda_3-\lambda_1)(\lambda_4-\lambda_1)}\frac{1}{\lambda_1+\mu^\nu}+\right. \\
+\frac{1}{(\lambda_1-\lambda_2)(\lambda_3-\lambda_2)(\lambda_4-\lambda_2)}\frac{1}{\lambda_2+\mu^\nu}+ \\
+\frac{1}{(\lambda_1-\lambda_3)(\lambda_2-\lambda_3)(\lambda_4-\lambda_3)}\frac{1}{\lambda_3+\mu^\nu}+ \\
\left.+\frac{1}{(\lambda_1-\lambda_4)(\lambda_2-\lambda_4)(\lambda_3-\lambda_4)}\frac{1}{\lambda_4+\mu^\nu}\right]
\end{aligned} \tag{5.190}$$

so that by inversion, the state probability $p_4(t)$ writes

$$p_4(t) = \prod_{j=1}^{3}\lambda_j \times \sum_{h=1}^{4}\frac{1}{\prod_{m=1,m\neq h}^{4}(\lambda_m-\lambda_h)}E_{\nu,1}(-\lambda_h t^\nu) \tag{5.191}$$

By induction, assuming that the solution holds true for $k-1$, we check that it solves the Cauchy problem for any $k > 1$. This

[10] By means of the Vandermonde formula for determinants with entries in geometric progression.

means that the Cauchy problem reads

$$\begin{cases} \frac{d^\nu}{dt^\nu} p_k^\nu(t) = \\ = -\lambda_k p_k^\nu(t) + \lambda_{k-1} \cdot \prod_{j=1}^{k-2} \lambda_j \sum_{h=1}^{k-1} \frac{1}{\prod_{h=1,m\neq h}^{k-1}(\lambda_m - \lambda_h)} E_{\nu,1}(-\lambda_h t^\nu), \\ k > 1 \\ p_k^\nu(0) = 0 \end{cases} \tag{5.192}$$

and its Laplace transform, after elementary manipulations, then reads

$$\mathcal{L}_k(\mu) = \\ = \prod_{j=1}^{k-1} \lambda_j \left[\sum_{h=1}^{k-1} \frac{1}{\prod_{m=1,m\neq h}^{k}(\lambda_m - \lambda_h)} \frac{\mu^{\nu-1}}{\mu^\nu + \lambda_h} + \right. \\ \left. - \frac{\mu^{\nu-1}}{\mu^\nu + \lambda_k} \sum_{h=1}^{k-1} \frac{1}{\prod_{m=1,m\neq h}^{k}(\lambda_m - \lambda_h)} \right] \tag{5.193}$$

It must be proved that the last term can be added into the summation, i.e. that

$$-\sum_{h=1}^{k-1} \frac{1}{\prod_{m=1,m\neq h}^{k}(\lambda_m - \lambda_h)} = \prod_{m=1,m\neq h}^{k} \frac{1}{(\lambda_m - \lambda_h)} \tag{5.194}$$

holds true. See Orsingher and Polito (2010) for further details on the technique used for manipulating the determinants arising from the algebra of the Laplace transform $\mathcal{L}_k$.

Once this is proved, the Laplace transform then reads

$$\mathcal{L}_k(\mu) = \prod_{j=1}^{k-1} \lambda_j \sum_{h=1}^{k} \frac{1}{\prod_{h=1,m\neq h}^{k}(\lambda_m - \lambda_h)} \cdot \frac{\mu^{\nu-1}}{\lambda_h + \mu^\nu} \tag{5.195}$$

from which it is now evident that, by inversion of the Laplace transform, the stated results holds true. □

The state probabilities p_k^ν are non-negative for $k \geq 1$, $t > 0$ and sum up to unity (see Remark 3.2 in Orsingher and Polito, 2010).

The process is non-homogeneous in time and its main qualitative feature is that at the beginning of the evolution, the very first offspring is more and more probable as the ν parameter decreases, as illustrated by the relation (see Remark 3.8 and eq. (3.59) in Orsingher and Polito, 2010)

$$\Pr\{N_\nu(dt) = n_0 + 1 \mid N_\nu(0) = n_0\} \sim \frac{\lambda n_0 (dt)^\nu}{\Gamma(\nu + 1)} \tag{5.196}$$

The linear process. The *fractional* linear birth process generalizes the classical Yule-Furry process where $\lambda_k = \lambda k$, $k \geq 1$, $\lambda > 0$.

The *fractional* equation for the state probabilities is

$$\frac{d^\nu p_k^\nu(t)}{dt^\nu} = -\lambda k p_k^\nu(t) + \lambda(k-1)p_{k-1}^\nu(t) \qquad 0 < \nu < 1, k \geq 1,$$
$$p_k(0) = \begin{cases} 1 & k = 1 \\ 0 & k \geq 2 \end{cases} \tag{5.197}$$

Orsingher and Polito (2010) provide the size distribution for the *linear* process both for a single progenitor and multiple progenitors. In the case of one single progenitor ($n_0 = 1$), it holds the following

Theorem 5.9. *The distribution of the fractional linear birth process is*

$$\begin{aligned} p_k^\nu(t) &= Pr\{N_\nu(t) = k \mid N_\nu(0) = 1\} = \\ &= \sum_{j-1}^{k} \binom{k-1}{j-1} (-1)^{j-1} E_{\nu,1}(-\lambda j t^\nu) \end{aligned} \tag{5.198}$$

Proof. We prove (5.198) by induction by solving the Cauchy problem

$$\frac{d^\nu p_k(t)}{dt^\nu} = -\lambda k p_k^\nu(t) + \lambda(k-1) \cdot \sum_{j=1}^{k-2} \binom{k-2}{j-1} (-1)^{j-1} E_{\nu,1}(-\lambda j t^\nu) \tag{5.199}$$

By applying the Laplace transform to both members of (5.199) one obtains

$$\mathcal{L}_{k,\nu}(\mu) = \lambda(k-1) \left\{ \sum_{j=1}^{k-1} \binom{k-2}{j-1} (-1)^{j-1} \frac{\mu^{\nu-1}}{\mu^\nu + \lambda j} \right\} \cdot \frac{1}{\mu^\nu + \lambda k} \tag{5.200}$$

For our convenience we write (5.200) as follows

$$\begin{aligned} &\mathcal{L}_{k,\nu}(\mu) = \\ &= \mu^{\nu-1} \cdot \left\{ \left[\frac{1}{\mu^\nu + \lambda} - \frac{1}{\mu^\nu + \lambda k} \right] - (k-1) \left[\frac{1}{\mu^\nu + 2\lambda} - \frac{1}{\mu^\nu + \lambda k} \right] + \right. \\ &\quad + \frac{(k-1)(k-2)}{2} \left[\frac{1}{\mu^\nu + 3\lambda} - \frac{1}{\mu^\nu + \lambda k} \right] + \dots \\ &\quad \left. + (k-1)(-1)^{k-2} \left[\frac{1}{\mu^\nu + (k-1)\lambda} - \frac{1}{\mu^\nu + \lambda k} \right] \right\} = \\ &= \mu^{\nu-1} \sum_{j=1}^{k-1} \binom{k-1}{j-1} (-1)^{j-1} \frac{1}{\mu^\nu + j\lambda} + \\ &\quad - \frac{\mu^{\nu-1}}{\mu^\nu + \lambda k} \cdot \sum_{j=1}^{k-1} \binom{k-1}{j-1} (-1)^{j-1} = \\ &= \sum_{j=1}^{k-1} \binom{k-1}{j-1} (-1)^{j-1} \frac{\mu^{\nu-1}}{\mu^\nu + j\lambda} \end{aligned} \tag{5.201}$$

and by inverting the Laplace transform (5.201) we obtain

$$\sum_{j=1}^{k-1} \binom{k-1}{j-1} (-1)^{j-1} E_{\nu,1}\left(-\lambda t^\nu\right) \tag{5.202}$$

and this confirms (5.198). □

In synthesis, the distribution in the case of $n_0 = 1$ and $n_0 > 1$ progenitors reads

$$p_k^\nu(t) =$$

$$\begin{cases} \binom{k-1}{j-1} \sum\limits_{j=1}^{k} (-1)^{j-1} E_{\nu,1}(-\lambda j t^\nu) & n_0 = 1 \\ \quad \text{(cit. Theor. 3.1)} & \\ \binom{n_0+k-1}{k} \sum\limits_{h=1}^{k} \binom{k}{h} (-1)^h E_{\nu,1}(-(n_0+h)\lambda t^\nu) & n_0 > 1 \\ \quad \text{(cit. Remark 3.8)} & \end{cases} \tag{5.203}$$

The following theorem proves, in the case of one progenitor, that the probabilities of the above distribution sums up to 1.

Theorem 5.10. *The probabilities* $p_k^\nu(t)$, $k \geq 1$, *of the distribution* (5.198) *sum up to* 1, *i.e.*

$$\sum_{k=1}^{\infty} p_k^\nu(t) = \sum_{k=1}^{\infty} \sum_{j=1}^{k} \binom{k-1}{j-1} (-1)^{j-1} E_{\nu,1}(-\lambda j t^\nu) = 1 \tag{5.204}$$

Proof. In the space of the Laplace transform equation (5.204) becomes

$$\begin{aligned} \sum_{k=1}^{\infty} \mathcal{L}_{\nu,k}(\mu) &= \\ = \sum_{k=1}^{\infty} \sum_{j=1}^{k} \binom{k-1}{j-1} (-1)^{j-1} \frac{\mu^{\nu-1}}{\mu^\nu + \lambda j} &= \\ = \frac{\mu^{\nu-1}}{\lambda} \sum_{k=1}^{\infty} \sum_{j=0}^{k-1} \binom{k-1}{j} (-1)^j \frac{1}{\frac{\mu^\nu}{\lambda} + j + 1} & \end{aligned} \tag{5.205}$$

Recognizing in equation the structure of an alternatig sum (i.e. sums of alternating sign, see Kirschenhofer, 1996) of the type $\sum_{k=0}^{N} (-1)^k \binom{N}{k} f(k)$, that, when $f(k) = \frac{1}{x+k}$, becomes the well-known formula

$$\sum_{k=0}^{N}\binom{N}{k}(-1)^k\frac{1}{x+k}=\frac{N!}{x(x+1)...(x+N)} \tag{5.206}$$

then, for $x=\frac{\mu^\nu}{\lambda}+1$, equation (5.6.1) simplifies into

$$\begin{aligned}
&\frac{\mu^{\nu-1}}{\lambda}\sum_{k=1}^{\infty}\frac{(k-1)!}{\left(\frac{\mu^\nu}{\lambda}+1\right)\left(\frac{\mu^\nu}{\lambda}+2\right)...\left(\frac{\mu^\nu}{\lambda}+k\right)}=\\
&=\frac{\mu^{\nu-1}}{\lambda}\sum_{h=0}^{\infty}\frac{\Gamma(h+1)\Gamma(\frac{\mu^\nu}{\lambda}+1)}{\Gamma(\frac{\mu^\nu}{\lambda}+(h+1)+1)}=\\
&=\frac{\mu^{\nu-1}}{\lambda}\sum_{h=0}^{\infty}\mathcal{B}(h+1,\frac{\mu^\nu}{\lambda}+1)=\\
&=\frac{\mu^{\nu-1}}{\lambda}\int_0^1\sum_{h=0}^{\infty}x^h(1-x)^{\frac{\mu^\nu}{\lambda}}dx=\\
&=\frac{\mu^{\nu-1}}{\lambda}\int_0^1(1-x)^{\frac{\mu^\nu}{\lambda}-1}dx=\\
&=\frac{\mu^{\nu-1}}{\lambda}\cdot\int_0^1 y^{\frac{\mu^\nu}{\lambda}-1}dy=\frac{1}{\mu}
\end{aligned} \tag{5.207}$$

□

which confirms the result in the Laplace transform space.

The moments. The mean value $\mathbb{E}[N^\nu(t)]$ can be obtained from the equation

$$\frac{\partial^\nu}{\partial t^\nu}G_\nu(u,t)=\lambda u(u-1)\frac{\partial}{\partial u}G_\nu(u,t) \tag{5.208}$$

by deriving both members w.r.t. u:

$$\frac{\partial^\nu}{\partial t^\nu}\frac{\partial}{\partial u}G_\nu(u,t)=\lambda[2u-1]\frac{\partial}{\partial u}G_\nu(u,t)+\lambda u(u-1)\frac{\partial^2}{\partial u}G_\nu(u,t) \tag{5.209}$$

which for $u = 1$ yields

$$\frac{\partial^\nu}{\partial t^\nu}\mathbb{E}[N_\nu(t)] = \lambda \mathbb{E}[N_\nu(t)] \tag{5.210}$$

Thus

$$\mathbb{E}[N_\nu(t)] = E_{\nu,1}(\lambda t^\nu) \tag{5.211}$$

denoting an increasingly rapid expansion of the population as ν decreases (see Fig.1, in Orsingher and Polito, 2010).

The variance reads

$$\mathbb{V}ar[N^\nu(t)] = 2E_{\nu,1}(2\lambda t^\nu) - E_{\nu,1}(\lambda t^\nu) - E^2_{\nu,1}(\lambda t^\nu) \tag{5.212}$$

which specializes to the variance of the classical Yule-Furry process when $\nu = 1$, i.e. $\mathbb{V}ar[N_1(t)] = e^{\lambda t}(e^{\lambda t} - 1)$.

5.6.2 The Fractional Pure-death Process

This paragraph deals with three fractional pure-death processes, namely the *non-linear* $\mathcal{M}^\nu(t)$, the *linear* $M(t)^\nu$ and the *sublinear* $\mathfrak{M}^\nu(t)$ death processes.

The non-linear process $\mathcal{M}^\nu(t)$. We now consider the *fractional* non-linear death process $\mathcal{M}^\nu(t)$ which extends the *non-linear* death process dealt with in Chapter 3.

The governing equation of the state probabilities $p_k(t)$, $t > 0$, in the fractional case, it becomes

$$\frac{d^\nu p_k(t)}{dt^\nu} = -\mu_k p_k(t) + \mu_{k+1} p_{k+1}(t) \quad 0 < k < n_0 \tag{5.213}$$

where for $k = 0$ and $k = n_0$ we have that

$$\begin{cases} \frac{d^\nu p_0(t)}{dt^\nu} = \mu_1 p_1(t) \\ \frac{d^\nu p_{n_0}(t)}{dt^\nu} = -\mu_{n_0} p_{n_0}(t) \end{cases} \tag{5.214}$$

with the initial conditions

$$p_k(0) = \begin{cases} 1 & k = n_0 \\ 0 & 0 \leq k < n_0 \end{cases} \tag{5.215}$$

Immediately, we have that

$$p_{n_0(t)} = E_{\nu,1}\left(-\mu_{n_0} t^\nu\right) \tag{5.216}$$

For $k = n_0 - 1$ we must solve

$$\begin{cases} \frac{d^\nu p_{n_0-1}(t)}{dt^\nu} = \mu_{n_0} E_{\nu,1}(-\mu_{n_0} t^\nu) - \mu_{n_0-1} p_{n_0-1}(t) \\ p_{n_0-1}(0) = 0 \end{cases} \tag{5.217}$$

The Laplace transform of equation (5.217) yields

$$\begin{aligned} \mathcal{L}[p_{n_0-1}(t)](\gamma) &= \frac{\mu_{n_0}\gamma^{\nu-1}}{(\gamma^\nu + \mu_{n_0})(\gamma^\nu + \mu_{n_0-1})} = \\ &= \frac{\mu_{n_0}\gamma^{\nu-1}}{\mu_{n_0-1} - \mu_{n_0}} \left(\frac{1}{\gamma^\nu + \mu_{n_0}} - \frac{1}{\gamma^\nu + \mu_{n_0-1}} \right) \end{aligned} \tag{5.218}$$

and its inverse is

$$p^\nu_{n_0-1}(t) = \frac{\mu_{n_0}}{\mu_{n_0-1} - \mu_{n_0}} \left[E_{\nu,1}(-\mu_{n_0} t^\nu) - E_{\nu,1}(-\mu_{n_0-1} t^\nu) \right] \tag{5.219}$$

In general (see Orsingher, Polito and Sakhno, 2010, formulas 3.17, 3.28) the distribution of the *fractional non-linear* death process reads ($n_0 > 1$)

$$p^\nu_k(t) = \begin{cases} E_{\nu,1}(-\mu_{n_0} t^\nu) & k = n_0 \\ \prod\limits_{j=k+1}^{n_0} \mu_j \cdot \sum\limits_{m=k}^{n_0} \frac{E_{\nu,1}(-\mu_m t^\nu)}{\prod_{h=k h\neq m}^{n_0}(\mu_h - \mu_m)} & 0 < k < n_0 \\ 1 - \sum\limits_{m=1}^{n_0} \prod\limits_{h=1 h\neq m}^{n_0} \frac{\mu_h}{\mu_h - \mu_m} E_{\nu,1}(-\mu_m t^\nu) & k = 0 \end{cases} \tag{5.220}$$

The linear process $M^\nu(t)$. The probability distribution of the population size for the linear process simplifies to (see Theor. 1 in

Orsingher, Polito and Sakhno, 2010)

$$p_k^\nu(t) = \Pr\{M^\nu(t) = k\} = \\ = \binom{n_0}{k} \sum_{h=0}^{n_0-k} \binom{n_0-k}{h} (-1)^h E_{\nu,1}(-(k+h)\mu t^\nu) \tag{5.221}$$

where $0 \le k \le n_0$, $t > 0$ and as usual $\nu \in (0,1]$.

The expected value obtains from the p.g.f. (see eq. (2.21) in Orsingher, Polito and Sakhno, 2010) as

$$\mathbb{E}[M^\nu(t)] = n_0 \cdot E_{\nu,1}(-\mu t^\nu), \quad \nu \in (0,1], t > 0 \tag{5.222}$$

The sublinear process $\mathfrak{M}^\nu(t)$. The *sublinear* case (the non-fractional process $\mathfrak{M}(t)$) obtains when the *infinitesimal* death rate probabilities have the form

$$\Pr\{\mathfrak{M}(t, t+dt] = -1 \mid \mathfrak{M}(t) = k\} = \mu(n_0-k+1)dt+o(dt) \tag{5.223}$$

i.e. when the death rate *increases with decreasing population size*, given an initial number n_0 of individuals in the population.

In the fractional case, the state probabilities $\mathfrak{p}_k^\nu(t) = \Pr\{\mathfrak{M}(t) = k \mid \mathfrak{M}(0) = n_0\}$ $(0 \le k \le n_0)$ satisfy the equations

$$\begin{cases} \frac{d^\nu \mathfrak{p}_k^\nu(t)}{dt^\nu} = -\mu(n_0-k+1)\mathfrak{p}_k^\nu(t) + \mu(n_0-k)\mathfrak{p}_{k+1}^\nu(t), & 1 \le k \le n_0 \\ \frac{d^\nu \mathfrak{p}_0^\nu(t)}{dt^\nu} = \mu n_0 \mathfrak{p}_1^\nu(t) \\ \mathfrak{p}_k^\nu(0) = \\ \begin{cases} 1, & k = n_0, \\ 0, & 0 \le k < n_0 \end{cases} \end{cases} \tag{5.224}$$

The size distribution reads (see eq. (3.31) in Orsingher, Polito and Sakhno (2010))

$$\mathfrak{p}_k^\nu(t) = \sum_{h=0}^{n_0-k} \binom{n_0-k}{h} (-1)^h E_{\nu,1}(-(h+1)\mu t^\nu), \quad 1 \le k \le n_0 \tag{5.225}$$

The extinction probability $\mathfrak{p}_0^\nu(t)$ reads

$$\mathfrak{p}_0^\nu(t) = \sum_{h=0}^{n_0} \binom{n_0}{h} (-1)^h E_{\nu,1}(-h\mu t^\nu). \tag{5.226}$$

Note that the sum of the state probabilities (for $1 \leq k \leq n_0$) and the extinction probability sum to one, i.e. $\sum_{k=1}^{n_0} \mathfrak{p}_k^\nu(t) + \mathfrak{p}_0^\nu(t) = 1$.

5.6.3 The Fractional Birth-death Processes

In this paragraph, based on Orsingher and Polito (2011), we finally analyze the *linear* version of the fractional birth-and-death process. To start with, a fundamental subordination relationship is proved, which is relevant to all the following distributional results. Thus, by exploiting this subordination relationship, the Laplace transform of the distribution (including the extinction probability) is obtained and then inverted.

The fundamental subordination relationship. To begin with, a fundamental subordination relation is proved (see Theorem 2.1 in Orsingher and Polito, 2011).

Theorem 5.11. *The fractional linear birth-death process $N_\nu(t)$, $t > 0$ can be represented as*

$$N_\nu(t) \overset{d}{=} N(T_{2\nu}(t)) \tag{5.227}$$

where $N(t)$ is a linear birth process, $T_{2\nu}(t)$ is a random process whose one-dimensional distribution coincides with the folded solution of the fractional diffusion equation

$$\frac{\partial^{2\nu} q}{\partial t^{2\nu}} = \frac{\partial^2 q}{\partial x^2} \quad 0 < \nu \leq 1, x \in \mathcal{R}, t > 0 \tag{5.228}$$

subject to the initial conditions

$$\begin{cases} q(x,0) = \delta(x) & 0 < \nu \leq \frac{1}{2} \\ q_t(x,0) = 0 & \frac{1}{2} < \nu \leq 1 \end{cases} \tag{5.229}$$

Proof. The probability generating function of the fractional linear birth-and-death process $G_\nu(u,t)$ satisfies the p.d.e.

$$\begin{cases} \frac{\partial^\nu G_\nu(u,t)}{\partial t^\nu} = (\lambda u - \mu)(u-1)\frac{\partial}{\partial u} G_\nu(u,t) \\ G_\nu(u,0) = u \end{cases} \tag{5.230}$$

The Laplace transform $\tilde{G}_\nu(u,t) = \int_0^\infty e^{-zt} G_\nu(u,t)dt$ of (5.230) reads

$$z^\nu \tilde{G}_\nu(u,z) - z^{\nu-1}u = (\lambda u - \mu)(u-1)\frac{\partial}{\partial u}\tilde{G}_\nu(u,z) \tag{5.231}$$

If the relationship

$$N_\nu(t) = N(T_{2\nu}(t)) \tag{5.232}$$

holds, the Laplace transform of both members of (5.227) become

$$\begin{aligned} &\int_0^\infty e^{-zt}\mathbb{E}\left[u^{N_\nu(t)}\right] dt = \tilde{G}_\nu(u,z) = \\ &= \int_0^\infty e^{-zt}\left[\sum_{k=0}^\infty u^k \Pr\{N_\nu(t) = k\}\right] dt \end{aligned} \tag{5.233}$$

Furthermore

$$\begin{aligned} &\int_0^\infty e^{-zt}\mathbb{E}\left[u^{N(T_{2\nu}(t))}\right] dt = \\ &= \int_0^\infty e^{-zt}\left[\sum_{k=0}^\infty u^k \int_0^\infty \Pr\{N(s)=k\}\cdot \Pr\{T_{2\nu}(t)\in ds\}\right] dt = \\ &= \int_0^\infty e^{-zt}\left[\int_0^\infty G(u,s)\Pr\{T_{2\nu}(t)\in ds\}\right] dt = \\ &= \int_0^\infty G(u,s) z^{\nu-1} e^{-sz^\nu} ds \end{aligned} \tag{5.234}$$

The last step is justified by formula (3.3) in Beghin and Orsingher (2004), i.e.

$$\int_0^\infty e^{-zt}\Pr\{T_{2\nu}(t)\in ds\}dt = z^{\nu-1}e^{-sz^\nu}ds \tag{5.235}$$

suitably adapted, that is, with $c = 1$ and folded.

We now prove that (5.234) solves equation (5.231). By inserting (5.234) into (5.231) we have that

$$\begin{aligned} z^{\nu} z^{\nu-1} \int_0^{\infty} G(u,s) e^{-sz^{\nu}} ds - z^{\nu-1} u = \\ = (u-1)(\lambda u - \mu) z^{\nu-1} \int_0^{\infty} \frac{\partial}{\partial u} G(u,s) e^{-sz^{\nu}} ds \end{aligned} \tag{5.236}$$

Since

$$\frac{\partial G}{\partial s} = (\lambda u - \mu)(u-1) \frac{\partial G}{\partial u} \tag{5.237}$$

equation (5.236) becomes

$$\begin{aligned} z^{\nu} \int_0^{\infty} G(u,s) e^{-sz^{\nu}} ds - u = \int_0^{\infty} \frac{\partial}{\partial s} G(u,s) e^{-sz^{\nu}} ds = \\ = G(u,s) e^{-sz^{\nu}} \Big|_{s=0}^{s=\infty} + z^{\nu} \int_0^{\infty} G(u,s) e^{-sz^{\nu}} ds = \\ = -u + z^{\nu} \int_0^{\infty} G(u,s) e^{-sz^{\nu}} ds \end{aligned} \tag{5.238}$$

Thus

$$\int_0^{\infty} e^{-\lambda t} \mathbb{E}\left[u^{N_{\nu}(t)}\right] dt = \int_0^{\infty} e^{-\lambda t} \mathbb{E}\left[u^{N(T_{2\nu}(t))}\right] dt \tag{5.239}$$

and both members satisfy equation (5.231).

□

Theorem 5.12. ***(The extinction probability)**. For a fractional linear birth-death process with rates $\lambda > \mu$, the probability of extinction has the form*

$$p_0^{\nu}(t) = Pr\{N_{\nu}(t) = 0\} = \frac{\mu}{\lambda} - \frac{\lambda - \mu}{\lambda} \sum_{m=1}^{\infty} \left(\frac{\mu}{\lambda}\right)^m E_{\nu,1}(-m(\lambda-\mu)t^{\nu}) \tag{5.240}$$

for $t > 0$, $0 < \nu \leq 1$, and where $E_{\nu,1}(x)$ is the Mittag-Leffler function.

Proof. The *extinction probability* is

$$\begin{cases} \Pr\{N(t)=0\} = \frac{\mu-\mu e^{-(\lambda-\mu)t}}{\lambda-\mu e^{-(\lambda-\mu)t}} & \text{linear birth-death process} \\ \Pr\{N_\nu(t)=0\} = \int_0^\infty \frac{\mu-\mu e^{-(\lambda-\mu)s}}{\lambda-\mu e^{-(\lambda-\mu)s}} \cdot \Pr\{T_{2\nu}\in ds\} & \text{fractional process} \end{cases} \tag{5.241}$$

where the second relation derives from the subordination relation (5.227).

The Laplace transform of the extinction probability reads

$$\begin{aligned} &\int_0^\infty e^{-zt}\Pr\{N_\nu(t)=0\}dt = \\ &= \int_0^\infty \frac{\mu-\mu e^{-(\lambda-\mu)s}}{\lambda-\mu e^{-(\lambda-\mu)s}} \int_0^\infty e^{-zt}\Pr\{T_{2\nu}\in ds\} = \\ &= \int_0^\infty \frac{\mu-\mu e^{-(\lambda-\mu)s}}{\lambda-\mu e^{-(\lambda-\mu)s}} \cdot (z^{\nu-1}e^{-z^\nu})ds = \\ &= \frac{\mu}{\lambda}\int_0^\infty \frac{1-e^{-(\lambda-\mu)s}}{1-\frac{\mu}{\lambda}e^{-(\lambda-\mu)s}} \cdot z^{\nu-1}e^{-z^\nu}ds = \\ &= \frac{\mu}{\lambda}\int_0^\infty (1-e^{-(\lambda-\mu)s}) \cdot \sum_{m=0}^\infty \left(\frac{\mu}{\lambda}e^{-(\lambda-\mu)s}\right)^m \cdot z^{\nu-1}e^{-sz^\nu}ds = \\ &= \frac{\mu}{\lambda}\sum_{m=0}^\infty \left(\frac{\mu}{\lambda}\right)^m z^{\nu-1} \cdot \int_0^\infty \left\{e^{-m(\lambda-\mu)s-sz^\nu} - e^{-(m+1)(\lambda-\mu)s-sz^\nu}\right\}ds = \\ &= \frac{\mu}{\lambda}\sum_{m=0}^\infty \left(\frac{\mu}{\lambda}\right)^m z^{\nu-1}\left[\frac{1}{(\lambda-\mu)m+z^\nu} - \frac{1}{(\lambda-\mu)(m+1)+z^\nu}\right] = \\ &= \frac{\mu}{\lambda}z^{\nu-1}\left[\sum_{m=0}^\infty \left(\frac{\mu}{\lambda}\right)^m \frac{1}{(\lambda-\mu)m+z^\nu} + \right. \\ &\left. -\sum_{m=1}^\infty \left(\frac{\mu}{\lambda}\right)^{m-1} \frac{1}{(\lambda-\mu)m+z^\nu}\right] = \\ &= \frac{\mu}{\lambda}z^{\nu-1}\left[\sum_{m=1}^\infty \frac{1}{(\lambda-\mu)m+z^\nu}\left(\frac{\mu}{\lambda}\right)^m\left(1-\frac{\lambda}{\mu}\right) + \frac{1}{z^\nu}\right] \end{aligned} \tag{5.242}$$

The inversion of the Laplace transforms in (5.242) yields the expression of the extinction probability

$$\begin{aligned} &\frac{\mu}{\lambda}\sum_{m=1}^{\infty}\left(\frac{\mu}{\lambda}\right)^m\left(1-\frac{\lambda}{\mu}\right)E_{\nu,1}(-(\lambda-\mu)mt^\nu)+\frac{\mu}{\lambda}= \\ &=\frac{\mu}{\lambda}-\frac{(\lambda-\mu)}{\lambda}\sum_{m=1}^{\infty}\left(\frac{\mu}{\lambda}\right)^m E_{\nu,1}(-(\lambda-\mu)mt^\nu) \end{aligned} \tag{5.243}$$

since the following Laplace transforms hold

$$\begin{cases} \int_0^\infty e^{-zt}\cdot E_{\nu,1}(-xt^\nu)dt=\frac{z^{\nu-1}}{z^\nu+x} \\ \int_0^\infty e^{-zt}\cdot 1dt=\frac{1}{z} \end{cases} \tag{5.244}$$

□

Theorem 5.13. ***(The distribution)***. *For the case $\lambda>\mu$, the state probabilities $p_k^\nu(t)$, $k\geq 1$, $t>0$, $0<\nu\leq 1$, in the fractional linear birth-death process $N_\nu(t)$, $t>0$, have the following form:*

$$\begin{aligned} p_k^\nu(t)=\left(\frac{\lambda-\mu}{\lambda}\right)^2\left(\sum_{h=0}^{\infty}\binom{k+h}{h}\left(\frac{\mu}{\lambda}\right)^h\sum_{m=0}^{k-1}(-1)^m\binom{k-1}{m}\cdot\right. \\ \left.\cdot E_{\nu,1}(-(h+m+1)(\lambda-\mu)t^\nu)\right) \end{aligned} \tag{5.245}$$

Proof. The distribution $\{p_k^\nu(t), k\geq 1\}$ is derived as follows. We take profit of the subordination $N_\nu(t)=N(T_{2\nu}(t))$ again. Hence

$$\begin{aligned} p_k^\nu(t)&=\int_0^\infty \Pr\{N(s)=k\}\cdot\Pr\{T_{2\nu}(t)\in ds\}= \\ &=\int_0^\infty(\lambda-\mu)^2\frac{e^{-(\lambda-\mu)s}\lambda^{k-1}(1-e^{-(\lambda-\mu)s})^{k-1}}{(\lambda-\mu e^{-(\lambda-\mu)s})^{k+1}}\cdot\Pr\{T_{2\nu}(t)\in ds\} \end{aligned} \tag{5.246}$$

Thus the Laplace transform of (5.246) yields

$$
\begin{aligned}
\mathcal{L}[p_k^\nu(t)](z) = \\
&= \int_0^\infty e^{-zt} p_k^\nu(t) dt = \\
&= \int_0^\infty (\lambda-\mu)^2 \frac{e^{-(\lambda-\mu)s}\lambda^{k-1}(1-e^{-(\lambda-\mu)s})^{k-1}}{(\lambda-\mu e^{-(\lambda-\mu)s})^{k+1}} \times \\
&\times \int_0^\infty e^{-zt} \Pr\{T_{2\nu}(t) \in ds\} dt = \\
&= (\lambda-\mu)^2 \lambda^{k-1} \cdot \\
&\cdot \int_0^\infty \frac{e^{-(\lambda-\mu)s}(1-e^{-(\lambda-\mu)s})^{k-1} \cdot (z^{\nu-1}e^{-sz^\nu})}{(\lambda-\mu e^{-(\lambda-\mu)s})^{k+1}} ds = \\
&= (\lambda-\mu)^2 \frac{\lambda^{k-1}}{\lambda^{k+1}} \cdot \\
&\cdot \int_0^\infty \sum_{h=0}^\infty \sum_{m=0}^{k-1} \binom{-(k+1)}{h} \left(-\frac{\mu}{\lambda} e^{-(\lambda-\mu)s}\right)^h \times \\
&\times \binom{k-1}{m} (-e^{-(\lambda-\mu)s})^m \cdot e^{-(\lambda-\mu)s} z^{\nu-1} e^{-sz^\nu} ds = \\
&= (\lambda-\mu)^2 \frac{\lambda^{k-1}}{\lambda^{k+1}} \cdot \\
&\cdot \sum_{h=0}^\infty \sum_{m=0}^{k-1} \binom{-(k+1)}{h} \left(-\frac{\mu}{\lambda}\right)^h \binom{k-1}{m} (-1)^m z^{\nu-1} \cdot \\
&\cdot \int_0^\infty \exp\{-sz^\nu - (\lambda-\mu)s - (h+m)(\lambda-\mu)s\} ds = \\
&= \left(\frac{\lambda-\mu}{\lambda}\right)^2 \cdot \\
&\cdot \sum_{h=0}^\infty \sum_{m=0}^{k-1} (-1)^h \binom{k+h}{h} \left(-\frac{\mu}{\lambda}\right)^h \binom{k-1}{m} (-1)^m z^{\nu-1} \cdot \\
&\cdot \int_0^\infty \exp\{-sz^\nu - (h+m+1)(\lambda-\mu)s\} = \\
&= \left(\frac{\lambda-\mu}{\lambda}\right)^2 \sum_{h=0}^\infty \sum_{m=0}^{k-1} \binom{k+h}{h} \binom{k-1}{m} (-1)^m \left(\frac{\mu}{\lambda}\right)^h \cdot \\
&\cdot \frac{z^{\nu-1}}{z^\nu + (h+m+1)(\lambda-\mu)}
\end{aligned} \tag{5.247}
$$

The inverse Laplace transform of (5.247) is

$$p_k^\nu(t) = \left(\frac{\lambda-\mu}{\lambda}\right)^2 \sum_{h=0}^{\infty}\sum_{m=0}^{k-1}\binom{k+h}{h}\binom{k-1}{m}(-1)^m\left(\frac{\mu}{\lambda}\right)^h \times$$
$$\times E_{\nu,1}(-(\lambda-\mu)(h+m+1)t^\nu) \tag{5.248}$$

□

The moments. We derive the first two factorial moments, as usual, by taking the first two derivatives of the p.g.f. in the unit argument and then combine them to obtain the variance.

As for the first moment, since

$$\mathbb{E}[N_\nu(t)] = \frac{dG_\nu(u,t)}{du}_{|u=1} \tag{5.249}$$

and given that $G_\nu(u,t)$ solves

$$\frac{d^\nu}{dt^\nu}G_\nu(u,t) = (\lambda u-\mu)(u-1)\frac{\partial G}{\partial u}G_\nu(u,t) \qquad |\, u \,|\leq 1 \tag{5.250}$$

then, taking the derivative w.r.t. u in (5.250), it also holds that

$$\frac{d^\nu}{dt^\nu}\left\{\frac{\partial G_\nu(u,t)}{\partial u}\right\} =$$
$$= [\lambda(u-1)+(\lambda u-\mu)]\frac{\partial G_\nu(u,t)}{\partial u} + (\lambda u-\mu)(u-1)\frac{\partial^2 G_\nu(u,t)}{\partial u^2} \tag{5.251}$$

and, at $u=1$, one obtains that the mean $E[N_\nu(t)]$ solves

$$\begin{cases} \frac{d^\nu}{dt^\nu}\mathbb{E}[N_\nu(t)] = (\lambda-\mu)\mathbb{E}[N_\nu(t)] \\ \mathbb{E}[N_\nu(0)] = 1 \end{cases} \tag{5.252}$$

whose solution is

$$\mathbb{E}[N_\nu(t)] = E_{\nu,1}((\lambda-\mu)t^\nu) \tag{5.253}$$

Remark 5.8.
Note that, for $\lambda > \mu$, the expected value behaves as (coincides with) that of a fractional linear *pure-birth* process with rate $\lambda - \mu > 0$.

An alternative derivation exploits the subordination relationship (5.227), i.e.

$$\begin{aligned} \mathbb{E}[N_\nu(t)] = \sum_{k=0}^{\infty} k\Pr\{N_\nu(t) = k\} = \\ = \left\{ \sum_{k=0}^{\infty} k \int_0^\infty \Pr\{N(s) = k\} \right\} \cdot \Pr\{T_{2\nu}(t) \in ds\} = \\ = \int_0^\infty e^{-(\lambda-\mu)s} \cdot \Pr\{T_{2\nu}(t) \in ds\} \end{aligned} \tag{5.254}$$

Hence, the Laplace transform of the mean function then reads

$$\begin{aligned} \int_0^\infty e^{-zt}\mathbb{E}[N_\nu(t)]dt = \\ = \int_0^\infty e^{-(\lambda-\mu)s} \cdot \int_0^\infty e^{-zt}\Pr\{T_{2\nu}(t) \in ds\}dt = \\ = \int_0^\infty e^{-(\lambda-\mu)s} \cdot (z^{\nu-1}e^{-sz^\nu})ds = \\ = \frac{z^{\nu-1}}{z^\nu - (\lambda-\mu)} = \int_0^\infty e^{-zt} E_{\nu,1}((\lambda-\mu)t^\nu)dz \end{aligned} \tag{5.255}$$

The second factorial moment. Taking the second derivative (w.r.t. u) in equation (5.250) yields the second factorial moment, i.e.

$$\begin{aligned} \frac{\partial^\nu}{\partial t^\nu}\frac{\partial^2 G_\nu(u,t)}{\partial u^2} = \frac{\partial^2}{\partial u^2}\left[(\lambda u - \mu)(u-1)\frac{\partial G_\nu(u,t)}{\partial u}\right] = \\ = \left[\frac{\partial^2}{\partial u^2}(\lambda u - \mu)(u-1)\right]\frac{\partial G_\nu(u,t)}{\partial u} + \\ +2\left[\frac{\partial}{\partial u}(\lambda u - \mu)(u-1)\frac{\partial^2 G_\nu(u,t)}{\partial u^2}\right] + \\ +(\lambda u - \mu)(u-1)\frac{\partial^3 G_\nu(u,t)}{\partial u^3} \end{aligned} \tag{5.256}$$

Then

$$\begin{cases} \frac{\partial^\nu}{\partial t^\nu}\mu_2(t) = 2\lambda \mathbb{E}N_\nu(t) + 2(\lambda-\mu)\mu_2(t) \\ \mu_2(0) = 0 \end{cases} \tag{5.257}$$

The, applying the Laplace transform to both sides, yields

$$z^\nu \int_0^\infty e^{-zt}\mu_2(t)dt = 2\lambda \int_0^\infty e^{-zt}\mu_1(t)dt + 2(\lambda-\mu)\int_0^\infty e^{-zt}\mu_2(t)dt \tag{5.258}$$

that is

$$\int_0^\infty e^{-zt}\mu_2(t)dt = 2\lambda \frac{z^{\nu-1}}{z^\nu - (\lambda-\mu)} \cdot \frac{1}{z^\nu - 2(\lambda-\mu)} \tag{5.259}$$

and inverting the Laplace transform leads to the explicit form of the second factorial moment as

$$\mu_2(t) = \frac{2\lambda}{\lambda-\mu}[E_{\nu,1}(2(\lambda-\mu)t^\nu) - E_{\nu,1}((\lambda-\mu)t^\nu)] \tag{5.260}$$

Thus, the first two factorial moments yields the following expression for the variance

$$\mathbb{V}ar[N_\nu(t)] =$$

$$= \frac{2\lambda}{\lambda-\mu}E_{\nu,1}(2(\lambda-\mu)t^\nu) - \frac{\lambda+\mu}{\lambda-\mu}E_{\nu,1}((\lambda-\mu)t^\nu) - E^2_{\nu,1}((\lambda-\mu)t^\nu) \tag{5.261}$$

Appendices

Appendix A1. The Probability Generating Function

The probability generating function (p.g.f.) of a discrete non negative r.v. $X : \{k, p_k; k = 0, 1, 2, ...\}$ is defined as

$$G_X(u) = \sum_{k=0}^{\infty} p_k u^k \quad |u| \leq 1 \qquad \text{for absolute convergence} \tag{A-262}$$

The name comes from the property

$$\frac{1}{k!}\frac{\partial^k G_X(u)}{\partial u^k}|_{u=0} = p_k \tag{A-263}$$

Consequently, if $G_X(u) = G_Y(u)$ then $p_X = p_Y$, i.e. if X and Y have the same p.g.f. then they have the same distribution.
Note that

$$\begin{cases} G_X(1) = \sum_{k=0}^{\infty} p_k = 1 \\ G_X^{(1)}(1) = \frac{\partial G_X(u)}{\partial u}_{|u=1} = bbE[X] \end{cases} \tag{A-264}$$

The k-th factorial moment is

$$\mu_{(k)} \equiv bbE[X(X-1)...(X-k+1)] = bbE\left[\frac{X!}{(X-k)!}\right] = \\ = \frac{\partial^k}{\partial u^k} G_X(u)_{|u=1} = G_X^{(k)}(1) \tag{A-265}$$

Therefore, the variance is

$$\begin{aligned} bbVar[X] &= bbE[X(X-1)] + bbE[X] - (bbE[X])^2 \\ &= \mu_{(2)}^X + \mu_{(1)}^X - (\mu_{(1)}^X)^2 \\ &= G_X^{(2)}(1) + G_X^{(1)}(1) - (G_X^{(1)}(1))^2. \end{aligned} \tag{A-266}$$

In general, the p.g.f. of a r.v. X is defined as

$$G_X(u) = bbE[u^X] \quad |u| \leq 1 \tag{A-267}$$

Given that the moment generating function (m.g.f.) is

$$M_X(\theta) \equiv bbE[e^{\theta X}] \tag{A-268}$$

then

$$M_X(\theta) = G_X(e^{\theta}) \tag{A-269}$$

If X_i, $i = 1, 2, ..., n$, are n independent r.v.'s and $S_n = \sum_{i=1}^{n} a_i X_i$, a_i constants, then

$$\begin{aligned} G_{S_n}(u) &= bbE[u^{S_n}] = bbE[u^{\sum_i a_i X_i}] = bbE[u^{a_1 X_1} ... u^{a_n X_n}] \\ &= \prod_{i=1}^{n} bbE[u^{a_i X_i}] = \prod_{i=1}^{n} G_{X_i}(u^{a_i}) \end{aligned} \tag{A-270}$$

For the special case $S = X_1 - X_2$

$$G_S(u) = G_{X_1}(u) G_{X_2}(\frac{1}{u}) \tag{A-271}$$

If N is a discrete r.v. independent of X_i and $S_N = \sum_{i=1}^{N} X_i$

$$G_{S_N}(u) = bbE\left[u^{\sum_i X_i}\right] = bbE_N\left[bbE\left[u^{\sum_{i=1} X_i} | N\right]\right]$$

$$= bbE_N\left[\prod_{i=1}^{N} bbE\left[u^{X_i}\right]\right] = bbE_N\left[\prod_{i=1}^{N} G_{X_i}(u)\right] = \sum_{n=1}^{\infty} p_n \prod_{i=1}^{n} G_{X_i}(u) \tag{A-272}$$

and if the X_i are also identically distributed (i.i.d.) then

$$G_{S_N(u)} = bbE_N[G_X(u)^N] = G_N(G_X(u)). \tag{A-273}$$

Examples

- The p.g.f. of a constant c is

$$G_c(u) = u^c \tag{A-274}$$

and the m.g.f. is

$$M_c(\theta) = e^{\theta c} \tag{A-275}$$

• The p.g.f. of a Bernoulli r.v. X with parameter p is

$$G_X(u) = bbE[u^X] = [pu^1 + (1-p)u^0] = [up + (1-p)]; \tag{A-276}$$

and the m.g.f. is

$$M_X(\theta) = pe^{\theta} + (1-p)e^0 = pe^{\theta} + 1 - p \tag{A-277}$$

• The p.g.f. of a Binomial r.v. $S_n = \sum_{i=1}^{n} X_i$ is

$$bbE[u^{S_n}] = bbE\left[\prod_{i=1}^{n} u^{X_i}\right] = \prod_{i=1}^{n} bbE\left[u^{X_i}\right] = bbE\left[u^{X_1}\right]^n \tag{A-278}$$

respectively by independence and equi-distribution of the X_i so that finally

$$= [u^1 p + (1-p)u^0]^n = [up + (1-p)]^n \tag{A-279}$$

where p is the parameter of the Bernoulli r.v. X_i.

The m.g.f. is

$$\begin{aligned} M_X(\theta) &= \sum_{k=0}^{n} e^{k\theta} \frac{n!}{k!(n-k)!} p^k (1-p)^{n-k} = \\ = \sum_{k=0}^{n} (pe^{\theta})^k &\frac{n!}{k!(n-k)!} (1-p)^{n-k} = (pe^{\theta} + 1 - p)^n \end{aligned} \tag{A-280}$$

• The p.g.f. of a Poisson(λ) r.v. is

$$G_X(u) = \sum_{k=0}^{\infty} e^{-\lambda} \frac{\lambda^k}{k!} u^k = e^{\lambda(u-1)}. \tag{A-281}$$

and the m.g.f. is

$$M_X(\theta) = e^{\lambda(e^{\theta}-1)} \tag{A-282}$$

Appendix B. Stable Distributions and Subordinators

B.1. Stable Random Variables

A random variable X is Stable iff $\forall\ a_1,\ a_2 \in bbR^+$ and $\forall X_1,\ X_2,\ X$, i.i.d., $\exists b > 0$ and $c \in bbR$ such that:

$$a_1 X_1 + a_2 X_2 \overset{cald}{\sim} bX + c \tag{B-283}$$

If X is Stable, $\exists \alpha \in (0, 2]$ called stability index, such that:

$$a_1^\alpha + a_2^\alpha = b^\alpha. \tag{B-284}$$

We write X is α-Stable.
The c.f. of an α-Stable r.v. X is the following:

$$bbE(e^{i\theta X}) = \begin{cases} e^{-\sigma^\alpha |\theta|^\alpha (1 - i\beta\theta \tan\frac{\alpha\pi}{2}) + i\mu\theta} & \alpha \neq 1 \\ e^{-\sigma|\theta|(1 + i\beta\frac{2}{\pi}\theta \ln|\theta|) + i\mu\theta} & \alpha = 1 \end{cases} \tag{B-285}$$

with $\sigma > 0$, $\beta \in [-1, 1]$, $\mu \in bbR$. We write also $X \sim S_\alpha(\sigma, \beta, \mu)$.
If $X > 0$ we have also the Laplace transform

$$bbE(e^{-\gamma X}) = \begin{cases} \exp\{-\frac{\sigma^\alpha \gamma^\alpha}{\cos\{\frac{\alpha\pi}{2}\}}\} & \alpha \neq 1 \\ \exp\{-\sigma\frac{2}{\pi}\ln\gamma\} & \alpha = 1 \end{cases} \tag{B-286}$$

Example. Let $X \sim N(\mu, \sigma^2)$, then $\alpha = 2$ and

$$b = \sqrt{a_1^2 + a_2^2}, \tag{B-287}$$

$$c = (a_1 + a_2 - b)\mu. \tag{B-288}$$

B.2. Stable Subordinators

The general idea of subordination refers to a process which evolves in an "operative" (as opposed to natural) random time.

- A subordinator is a (a.s.) non-decreasing right-continuous process with stationary, i.i.d. increments (a case of Levy process). A Stable subordinator is an α-Stable process $S_\alpha(t)$; an inverse Stable subordinator $L(x)$ represents the first passage time of a Stable subordinator S_α for a level x, i.e. $L(x) = \inf\{t : S_\alpha(t) \geq x\}$, $x > 0$. The following results hold.

- α-Stable subordinators $S_\alpha(t)$, $0 < \alpha < 1$, are characterized by Laplace transform of the form

$$bbE[e^{-\theta S_\alpha(t)}] = e^{-\theta^\alpha t} \qquad \theta > 0, t > 0 \tag{B-289}$$

- α-Stable subordinators, $0 < \alpha < 1$, scale in time according to a power-law with exponent greater than unity, i.e.

$$S_\alpha(ct) \sim c^{\frac{1}{\alpha}} S_\alpha(t), \qquad c > 0 \tag{B-290}$$

  Inverse α-Stable subordinators, $0 < \alpha < 1$, scale in time according to a power-law with an exponent smaller than unity, i.e.

$$S_\alpha(ct) \overset{d}{\sim} c^\alpha S_\alpha(t), \qquad c > 0 \tag{B-291}$$

  The density function of an $\alpha-$Stable subordinator is given by

$$h_\alpha(x,t) = \alpha t x^{-(\alpha+1)} W_\alpha(t x^{-\alpha}) \qquad x > 0 \tag{B-292}$$

  while the density function of its inverse $\alpha-$Stable subordinator is given by

$$l_\alpha(t,x) = t^{-\alpha} W_\alpha(t^{-\alpha} x) \qquad x > 0 \tag{B-293}$$

  where $W_\alpha(x)$ is the Wright function

$$W_\alpha(x) \equiv W_{-\alpha,1-\alpha}(x) = \sum_{k=0}^{\infty} \frac{(-x)^k}{k!\Gamma(-\alpha k + (1-\alpha))}. \tag{B-294}$$

  The two densities are related as

$$l_\alpha(t,x) = -\frac{\partial}{\partial t} \int_0^x h_\alpha(y,t)dy \tag{B-295}$$

Closed form densities arise for $\alpha = \frac{1}{2}$, in which case

$$h_{\frac{1}{2}}(x,t) = \frac{t}{\sqrt{4\pi x^3}} e^{-\frac{t^2}{4x}} \tag{B-296}$$

which corresponds to the first passage time distribution (Inverse Gaussian) of a standard Brownian motion through the *level* $\frac{t}{\sqrt{2}}$ (or, equivalently, of a Brownian motion with variance $\sqrt{2}$ through a level t), and

$$l_{\frac{1}{2}}(t,x) = \frac{1}{\sqrt{\pi x}} e^{-\frac{t^2}{4x}} \tag{B-297}$$

which corresponds to the probability distribution of the maximum of a Brownian motion over a *time* interval $(0, x)$.

Appendix C. Fractional Derivative and Integrals

The Riemann-Liouville fractional integral of order α is defined as

$$(I^\alpha f)(x) = \frac{1}{\Gamma(\alpha)} \int_0^x (x-y)^{\alpha-1} f(y) dy, \quad \alpha > 1 \tag{C-298}$$

The fractional derivative of order α of a (well-behaved) function is the ordinary m-th derivative D^m of a fractional Riemann-Liouville integral of order $m - \alpha$, i.e.

$$(D^\alpha f)(x) = [D^m (I^{m-\alpha} f)](x), \quad \text{for } m-1 < \alpha < m, \alpha > -1 \tag{C-299}$$

In fact the Riemann-Liouville fractional derivative of a function $f(x)$ is defined explicitly as

$$(D^\alpha f)(x) = \begin{cases} \frac{1}{\Gamma(m-\alpha)} \frac{d^m}{dx^m} \int_0^x \frac{f(y)}{(x-y)^{\alpha+1-m}} dy & m-1 < \alpha < m \\ \frac{d^m}{dx^m} f(x) & \alpha = m \end{cases} \tag{C-300}$$

Note that the fractional integral is an extension of the Cauchy formula:

$$\int_0^x dx_1 \int_0^{x_1} dx_2 ... \int_0^{x_{n-1}} f(x_n)dx_n = \frac{1}{(n-1)!}\int_0^x f(x_n)(x-x_n)^{n-1}dx_n. \tag{C-301}$$

and the m-th (ordinary) derivative of a power function x^n is

$$D^m x^n = \frac{d^m x^n}{dx^m} = n(n-1)(n-m+1)x^{n-m} = \frac{n!}{(n-m)!}x^{n-m} = \frac{\Gamma(n+1)}{\Gamma(n-m+1)}x^{n-m}, \qquad m \leq n. \tag{C-302}$$

By analogy, the fractional α-th derivative can be obtained directly as

$$\frac{d^\alpha}{dx^\alpha}x^n = \frac{\Gamma(n+1)}{\Gamma(n-\alpha+1)}x^{n-\alpha}, \alpha \leq n. \tag{C-303}$$

For example, the derivative of half order (i.e., $\alpha = \frac{1}{2}$) for $n = 1$ gives

$$\frac{d^{\frac{1}{2}}}{dx^{\frac{1}{2}}}x = \frac{\Gamma(2)}{\Gamma(\frac{3}{2})}x^{\frac{1}{2}} = \frac{x^{\frac{1}{2}}}{\Gamma(\frac{3}{2})} = \frac{1}{\frac{1}{2}\sqrt{\pi}}x^{\frac{1}{2}}. \tag{C-304}$$

The Dzherbashyan-Caputo fractional derivative is defined as

$$(D_c^\alpha f)(x) = \begin{cases} \frac{1}{\Gamma(m-\alpha)}\int_0^x \frac{\frac{d^m}{dy^m}f(y)}{(x-y)^{\alpha+1-m}}dy & m-1 < \alpha < m \\ \frac{d^m}{dx^m}f(x) & \alpha = m \end{cases} \tag{C-305}$$

Appendix D. Mittag-Leffler and Gamma functions

D.1 The Mittag-Leffler Function.

The Mittag-Leffler function was originally defined by Gosta Mittag-Leffler as a *one-parameter* power series on the complex plane; when restricted to a real function, its power series definition reads:

$$E_\alpha(x) = \sum_{k=0}^{\infty} \frac{x^k}{\Gamma(\alpha k + 1)} \tag{D-306}$$

where we limit ourselves to the real domain, i.e. $x \in bbR$, and $0 \leq \alpha \leq 1$.

Its Laplace transform $LT[E_\alpha(t)](\theta)$ has several properties, among which

$$LT[E_\alpha(\pm t^a)](\theta) = \int_0^\infty e^{\theta t} E_\alpha(\pm t^a) dt = \frac{\theta^{a-1}}{\theta^a \mp 1} \tag{D-307}$$

turns useful in applications to fractional processes.

Another relevant property of the Mittag-Leffler function of negative argument, $E_\alpha(-t)$, is its *complete monotonicity*, that is its (integer) derivatives $E_\alpha^{(n)}(\cdot)$ are alternating in sign, i.e.

$$(-1)^n E_\alpha^{(n)}(-t) \geq 0 \qquad \forall t \in (0, \infty), \quad 0 \leq \alpha < 1 \tag{D-308}$$

The (Riemann-Liouville) fractional integral of the Mittag-Leffler function "mimics" the "invariance behaviour" of the classical exponential funcion w.r.t. the Riemann integral, that is:

$$\frac{\lambda}{\Gamma(\alpha)} \int_0^x \frac{E_\alpha(\lambda t^\alpha)}{(x-t)^{1-\alpha}} dt = E_\alpha(\lambda x^\alpha) - 1 \tag{D-309}$$

D.2 the Gamma Function and Euler's Reflection Formula.

Euler's Gamma reflection property reads

$$\Gamma(x)\Gamma(1-x) = \frac{\pi}{\sin(\pi x)} \tag{D-310}$$

and relates the Gamma function to trigonometric functions through their respective zeros.

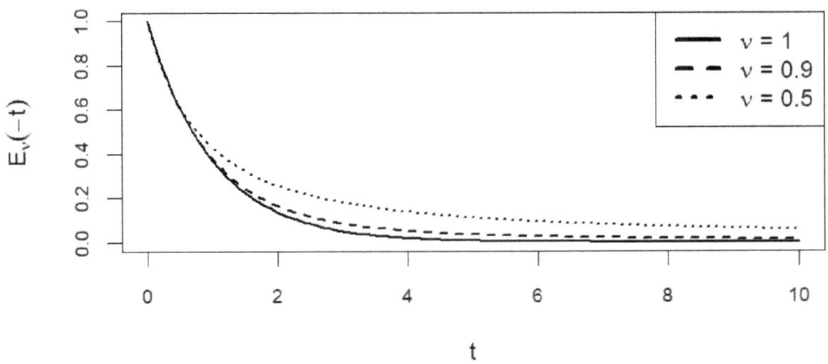

Fig. D.1. The (one parameter) Mittag-Leffler function.

Indeed, limiting ourselves to the real line - the zeros of the $\sin(\pi x)$ function, spanning the set of positive and negative integers, are "collected" by the product of the two Gamma functions: indeed, while the factor $\Gamma(x)$ "collects" the negative zeros of $\sin(\pi x)$, while the Gamma function "reflected" in the argument $1 - x$ "collects" its positive zeros. Note that the presence of the sine function at the denominator reflects the fact that the Gamma function has no zeros (but it has asymptotes to infinity at each negative integers and at zero, see Fig. D.2).

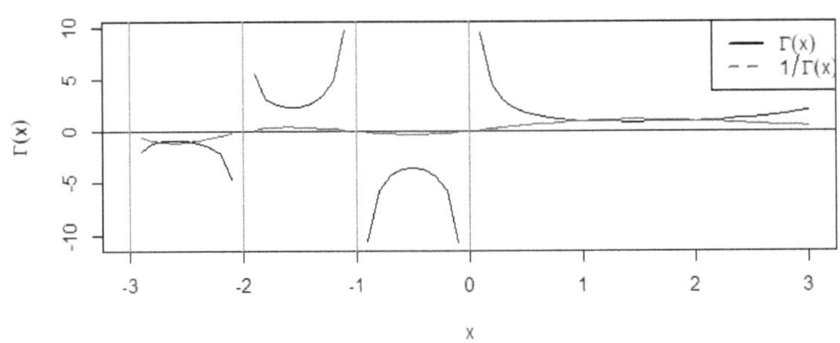

Fig. D.2. Gamma function and its reciprocal.

Appendix E. Tools for Data Analysis

E.1. QQ and PP Plots

Given a data set, the PP plots compare the *empirical* cumulative distribution function (CDF) with a specified *theoretical* cumulative distribution function $F(\cdot)$. A QQ plot compares the empirical quantiles with the quantiles of a *standardized* theoretical distribution.

Hence, PP plots require the estimation of the distribution parameters (e.g. scale and location), while QQ plots refer to standardized distributions and do not require them.

In both QQ and PP plots linearity indicates a good fit to the data. In QQ plots the intercept and slope correspond to the location and scale parameters of the theoretical distribution, while in PP plots the reference line is always the bisectrix.

Consequently, QQ plots are used to select a distribution within a family of distributions that vary only by scale and location, and hence help in fitting the tails of the empirical distribution.

Viceversa, PP plots help in discriminating the central part of the distribution, where the difference between the empirical and theoretical CDFs is larger.

E.2. The Chi-square Goodness-of-fit Test

The Pearson χ^2 test framework was introduced by Pearson (1900) and it has three standard applications:

- test of independence
- goodness-of-fit test
- distribution test

We focus on the goodness-of-fit test. The test statistic is a sum of normalized squared residuals (i.e. observed minus expected values) over grouped data, i.e.

$$X^2 = \sum_{k=1}^{m} \frac{(n_k - np_k)^2}{np_k} \tag{E-311}$$

where data are grouped into k (mutually exclusive) categories or classes (observations), n_k are the *empirical* (absolute) frequencies of each category (s.t. $\sum_{i=1}^{k} n_k = n$) and the p_k's are *theoretical* (relative) frequencies (probabilities) based on a *null* hypothesis H_0 about the data generator distribution. The *expected* value for each category is then the product np_k.

The test is said *simple* when the parameters are known ex-ante, and *composed* when the parameters are estimated *before running the test*, i.e.:

$$X^2 = \sum_{k=1}^{m} \frac{(n_k - np_k(\underline{\theta}))^2}{np_k(\underline{\theta})} \tag{E-312}$$

where $\underline{\theta}$ is the parameter vector of the *null* distribution and now the expected values $np_k(\underline{\theta})$ depend on the estimated parameters.

The null hypothesis H_0 then is rejected when the empirical realization of the test statistic X^2 falls in the rejection region of the (one-sided) test, i.e. when X^2 is greater (beyond) the critical value t^*, determined on the table of the χ^2 distribution (cf. Table A) upon specification of the number of degrees-of-freedom.

When the data generator distribution is fully known (included its parameters) or does not have parameters (e.g. Uniform distributions), then the *asymptotic* distribution of the test statistic X^2 is χ^2 with $m - 1$ degrees-of-freedom, i.e. $X^2 \sim_{n\to\infty} \chi^2(m-1)$.

However, if the parameters are estimated, then the asymptotic distribution reduces its degrees-of-freedom by the number p of estimated parameters (e.g. $p = 1$ for the Poisson distribution and $p = 2$ in the case of the Negative Binomial).

If the parameter estimates come from a *multinomial* ML estimation (i.e. based on grouped data), then the asymptotic distribution of the test statistic X^2 follows a $\chi^2(m-p-1)$ distribution.

Remark.
Other distributions arise when the parameters are estimated by *full* Maximum-Likelihood (i.e. on original observed, ungrouped data) or by other estimation methods (e.g. Method-of-Moments).

Note that if the parameters are estimated by means of MM estimators (or any other estimator the is not BAN, *Best Asymptotically Normal*), then the asymptotic distribution of the test statistic is not known and in general is even not a χ^2.

See Chernoff and Lehmann (1954) for more details on the correct asymptotic distributions of the test in these cases.

Albrecht (1980) resumes these results for actuarial applications.

Table A. Chi-Square probabilities $\Pr\{\chi^2 \geq q\}$

df \ q	0.99	0.975	0.95	0.90	0.10	0.05	0.025	0.01
1	—	0.001	0.004	0.016	2.706	3.84	5.02	6.64
2	0.020	0.051	0.103	0.211	4.605	5.99	7.38	9.21
3	0.115	0.216	0.352	0.584	6.251	7.82	9.35	11.35
4	0.297	0.484	0.711	1.064	7.78	9.45	11.143	13.28
5	0.554	0.831	1.145	1.610	9.24	11.07	12.833	15.09
6	0.872	1.237	1.635	2.204	10.65	12.56	14.449	16.81
7	1.239	1.690	2.167	2.833	12.02	14.07	16.013	18.48
8	1.646	2.180	2.733	3.490	13.37	15.51	17.535	20.09
9	2.088	2.700	3.325	4.168	14.68	16.92	19.023	21.67
10	2.558	3.247	3.940	4.865	15.99	18.31	20.483	23.21
11	3.053	3.816	4.575	5.578	17.28	19.68	21.920	24.73
12	3.571	4.404	5.226	6.304	18.55	21.03	23.337	26.22
13	4.107	5.009	5.892	7.042	19.81	22.37	24.736	27.69
14	4.660	5.629	6.571	7.790	21.06	23.69	26.119	29.14
15	5.229	6.262	7.261	8.547	22.31	24.99	27.488	30.58

E.3. The Kolmogorov–Smirnov Goodness-of-fit Test

The Kolmogorov–Smirnov test (KS test) is a nonparametric test of the equality of *continuous* probability distributions.

It is used to compare a sample with a reference probability distribution (one-sample K–S test).

The Kolmogorov–Smirnov statistic quantifies a distance between the empirical distribution function of the sample and the cumulative distribution function of the reference distribution: the test statistic for a given cumulative distribution function $F(x)$ is:

$$D_n = \sup_x \mid F_n(x) - F(x) \mid \tag{E-313}$$

where the supremum is taken over the set of point-wise distances. Intuitively, the statistic takes the largest absolute difference between the two distribution functions across all x values.

Depending on the specific distribution $F(x)$, specific tables are valid for the test (e.g. Normal and Log-Normal distributions form a single table as the logarithm can be applied to the crude data).

Appendix F. R Code of Some Examples and Applications

This appendix provides the R code backing some examples and applications of the book.

F.1. Poisson Distribution (Chapter 1, Fig. 1.1)

The code represents the Poisson distribution with three different parameter values λ by means of the *barplot()* function (see chapter 1, Fig. 1.1).

```
> my.names = c("k = 0", "k = 1", "k = 2", "k = 3",
  "k = 4")

> Poisson.1 = dpois(0:4, lambda = 1)

> Poisson.2 = dpois(0:4, lambda = 0.5)

> Poisson.3 = dpois(0:4, lambda = 0.1)

> dati = rbind(Poisson.3, Poisson.2, Poisson.1)

> my.bar = barplot(dati,
 beside = TRUE,
 names.arg=my.names,
 col = c("black", "darkgray", "lightgray"),
 ylab = expression(p[k]), # "frequencies",
 xlab = "k",
 ylim = c(0,1)
 )

> legend("topright", c("lambda = 0.1",
 "lambda = 0.5", "lambda = 1"),
 fill = c("black", "darkgray", "lightgray"))

> text(my.bar, dati + 0.05, labels = round(dati,3))
```

F.2. Pearson's Chi-square Tests (Chapter 1, Example I.1)

The codes reported below conduct two Pearson's chi-square tests on Von Bortkewitsch (1898) classical observations of the number of deaths due to horse kicks (in ten corps of the Prussian army between 1875 and 1894) with two different null hypotheses, namely the Poisson and the Negative Binomial (NB), respectively.

F.2.1. Pearson's Chi-square Test (Poisson Distribution)

The code conducts Pearson's chi-squared test for the null hypothesis that the observations are compatible with a Poisson distribution (see Chapter 1, Example I.1).

```
> NumberOfDeaths = 0:3

> print(NumberOfDeaths)

> actuals = c(109,65,22,3+1+0)

> print(actuals)

> TotalDeaths = as.numeric(NumberOfDeaths %*% actuals)

> print(TotalDeaths)

> NumberOfObs = sum(actuals)

> print(NumberOfObs)

> lambda = TotalDeaths/NumberOfObs

> print(lambda)

> predicted = c(dpois(0:2,lambda)*NumberOfObs,
  (1 - ppois(2,lambda))*NumberOfObs)

> print(predicted)

> table.1 = cbind(NumberOfDeaths,actuals)

> colnames(table.1)[2] = "Number of corps with x deaths
  in 20 years"

> require (knitr)

> require(kableExtra)
```

```
> require(dplyr)

> table.1 %>% kable()

> print(table.1)

> table.2 = cbind(table.1, predicted)

> colnames(table.2)[3] = "Poisson prediction"

> table.2 %>%  kable(full_width = F)

> barplot(t(table.2[,2:3]), axis.lty = 1,

    beside=T, col=c("grey","orange"),

    names.arg=NumberOfDeaths,

    xlab="k", ylab="frequencies",

    main = "Deaths from horse kicks in Prussian army corps,
  1875-1894",

    legend.text = c("empirical","Poisson")

    )

> chisq = chisq.test(actuals, p=predicted/sum(predicted),
  rescale.p = T)

> chisq$parameter = c(df=2)

> chisq$p.value = pchisq(chisq$statistic, df=chisq$parameter,
  lower.tail=FALSE)

> chisq
```

F.2.2. Pearson's Chi-square Test (Negative Binomial Distribution)

The code conducts Pearson's chi-squared test for the null hypothesis that the observations are compatible with a Negative Binomial distribution (see Chapter 1, Example I.1).

```
> NumberOfDeaths = 0:3

> print(NumberOfDeaths)

> actuals = c(109,65,22,3+1+0)

> print(actuals)

> TotalDeaths = as.numeric(NumberOfDeaths %*% actuals)

> print(TotalDeaths)

> NumberOfObs = 200

> lambda = 0.61

> print(lambda)

> require(actuar)

> tau = 9.484

> nu = 5.785

> p = tau/(1 + tau)

> negbin = c(dnbinom(0:2, size = nu, prob = p),

> 1 - pnbinom(2, size = nu, prob = p))

> predicted = negbin*NumberOfObs
```

```
> print(predicted)

> table.1 = cbind(NumberOfDeaths,actuals)

> colnames(table.1)[2] = "Number of corps with x deaths in
  20 years"

> require (knitr)

> require(kableExtra)

> require(dplyr)

> table.1 %>%

> kable()

> table.2 = cbind(table.1, predicted)

> colnames(table.2)[3]="NegBin prediction"

> table.2 %>%

> kable(full_width = F)

> barplot(t(table.2[,2:3]), axis.lty = 1,

    beside=T, col=c("grey","orange"),

    names.arg=NumberOfDeaths,

    xlab="k",

    ylab="frequencies",

    main = "Deaths from horse kicks in Prussian army
            corps,
```

```
1875-1894",

    legend.text = c("empirical","NegBin")

    )

> chisq = chisq.test(actuals, p=predicted/sum(predicted), )
          rescale.p = T)

> chisq$parameter = c(df=4-1-2)

> chisq$p.value = pchisq(chisq$statistic, df=chisq$parameter,
 lower.tail=FALSE)

> chisq
```

F.3. Compound Poisson (Chapter 1, § 1.5.4)

The code reported below builds on the R package *fitdistrplus* and related references in order to fit a Compound-Poisson Log-Normal distribution to Danish fire loss insurance data (see Chapter 1, § 1.5.4).

```
> require(fitdistrplus)

> my.danish = data("danishuni", "fitdistrplus")

> danish = data.frame(my.danish)

> str("danishuni")

> head(danish)

> tail(my.danish)

> View(my.danish)
```

```
> head(danishuni)

> tail(danishuni)

> my.csv = read.csv2(file = "danish.fire.csv")

> head(my.csv)

> my.df = data.frame(my.csv)

> View(my.df)

> my.dt = my.df$Date[-1] - my.df$Date[-2167]

> fdanish.ln.MLE = fitdist(danishuni$Loss, "lnorm")

> fdanish.ln.MME = fitdist(danishuni$Loss, "lnorm",
  method = "mme", order = 1:2)

> my.cpp = cdfcomp(list(fdanish.ln.MLE, fdanish.ln.MME),
 legend = c("LogNormal MLE", "LogNormal MME"),
 main = "", lwd = 2,
 xlogscale = TRUE,
 datapch = 20
 )

> require(actuar)

> fdanish.Pa.MLE = fitdist(danishuni$Loss, "pareto",
 start = c(shape = 10, scale = 10),
 lower = 2+1e-6, upper = Inf
 )

> fdanish.Pa.MME = fitdist(danishuni$Loss, "pareto",
 method = "mme", order = 1:2,
 memp = "memp", start = c(shape = 10, scale = 10),
 lower = c(2+1e-6, 2+1e-6), upper = c(Inf, Inf)
```

```
)

> my.cpp = cdfcomp(list(fdanish.Pa.MLE, fdanish.Pa.MME),
legend = c("lognormal MLE", "lognormal MME"),
main = "Fitting a lognormal distribution",
xlogscale = TRUE,
datapch = 20
)

> require(fitdistrplus)

> fw.expon = fitdist(data = my.df$Loss,
method = "mle", "exp")

> print(fw.expon)

> fw.weibull = fitdist(data = my.df$Loss,
method = "mle", "weibull")

> print(fw.weibull)

> fw.gamma = fitdist(data = my.df$Loss,
method = "mle", "gamma")

> print(fw.gamma)

> fw.lognorm = fitdist(data = my.df$Loss,
method = "mle", "lnorm")

> print(fw.lognorm)

> fw.pareto = fitdist(data = my.df$Loss,
method = "mme", "pareto",
order=c(1, 2),
memp="memp",
start=list(shape=10, scale=10),
lower=1,
```

```
upper=Inf
)

> print(fw.pareto)

> my.qq.expon = qqcomp(fw.expon,
main = "qq-plot: fit code")

> my.qq.weibull = qqcomp(fw.weibull,
main = "qq-plot: fit code")

> my.qq.gamma = qqcomp(fw.gamma,
main = "qq-plot: fit code")

> my.qq.lognorm = qqcomp(fw.lognorm,
main = "qq-plot: fit code")

> my.qq.pareto = qqcomp(fw.pareto,
main = "qq-plot: fit code")

> my.pp.weibull = ppcomp(fw,
main = "pp-plot: fit centro")

> my.qq.pareto = qqcomp(fw,
main = "qq-plot: fit code")

> par(mfrow=c(1,2))

> my.qq.weibull = qqcomp(fw, main = "qq-plot: Weibull")

> my.pp.weibull = ppcomp(fw, main = "pp-plot: Weibull")
```

F.4. Hawkes Processes (Chapter 1, § 1.9.2 Example and Fig. 1.16)

This section reports R code related to the estimation of the parameters of a Hawkes process $N^H(t)$ and to the plot of its intensity function $\tilde{\lambda}(t)$ (see Chapter 1, § 1.9.2, Example).

F.4.1. Hawkes Process (Parameter Estimation)

The code reported below is based on car accident data and it estimates the parameters of a Hawkes process (see Chapter 1, § 1.9.2, Example).

```
> getwd()

> setwd("D:/Dati/Profili/Documenti")

> my.csv.2019 = read.csv2(file =
 "OpenData.Incidenti.2019.csv", header = TRUE)

> my.df.2019 = data.frame(my.csv.2019)

> require(dplyr)

> my.df.0 = my.df.2019 %>% distinct(Protocollo,
 .keep_all = TRUE)

> d1 = as.POSIXlt(my.df.0$DataOraIncidente,
 format = "%d/%m/%Y %H:%M")

> head(d1)

> d2 = d1$hour

> head(d2)

> print(d2)

> is.na(d2)

> sum(is.na(d2))

> my.df.0[[38]] = d2
```

```
> my.df.1 = filter(my.df.0, d2 < 7)

> require(dplyr)

> csv.flaminia = read.csv2(file = "flaminia.csv",
header = TRUE)

> df.flaminia = data.frame(csv.flaminia)

> df.flaminia.night = filter(df.flaminia, as.numeric(V38) < 7)

> strip.flaminia.night =
strptime(df.flaminia.night$DataOraIncidente,
format = "%d/%m/%Y %H:%M")

> head(strip.flaminia.night)

> n.flaminia.night = length(strip.flaminia.night)

> diff.flaminia.night = difftime(strip.flaminia.night[-1],
strip.flaminia.night[-n.flaminia.night], units = "days")

> length(diff.flaminia.night)

> head(diff.flaminia.night)

> tail(diff.flaminia.night)

> m.flaminia.night = mean(diff.flaminia.night)

> print(m.flaminia.night)

> flaminia.night.points =
cumsum(as.numeric(diff.flaminia.night))/365

> head(flaminia.night.points)
```

```
> tail(flaminia.night.points)

> csv.appia = read.csv2(file = "appia.csv", header = TRUE)

> df.appia = data.frame(csv.appia)

> df.appia.night = filter(df.appia, as.numeric(V38) < 7)

> strip.appia.night =
strptime(df.appia.night$DataOraIncidente,
format = "%d/%m/%Y %H:%M")

> head(strip.appia.night)

> n.appia.night = length(strip.appia.night)

> diff.appia.night = difftime(strip.appia.night[-1],
strip.appia.night[-n.appia.night], units = "days")

> length(diff.appia.night)

> head(diff.appia.night)

> tail(diff.appia.night)

> m.appia.night = mean(diff.appia.night)

> print(m.appia.night)

> appia.night.points =
cumsum(as.numeric(diff.appia.night))/365

> head(appia.night.points)

> tail(appia.night.points)
```

```
> require(hawkesbow)

> max.flaminia.night = max(flaminia.night.points)

> x.end.flaminia.night = round(max.flaminia.night)

> print(x.end.flaminia.night)

> lik.flaminia.night = mle(flaminia.night.points,
  "Exponential", x.end.flaminia.night)

> summary(lik.flaminia.night)

> par.flaminia.night = lik.flaminia.night$par

> print(par.flaminia.night)

> max.appia.night = max(appia.night.points)

> x.end.appia.night = round(max.appia.night)

> print(x.end.appia.night)

> lik.appia.night = mle(appia.night.points, "Exponential",
  x.end.appia.night)

> summary(lik.appia.night)

> par.appia.night = lik.appia.night$par

> print(par.appia.night)
```

F.4.2. Hawkes Process (Intensity Function)

The code reported below implements the function of R package *hawkesbow* plotting a Hawkes process intensity (see Chapter 1, Fig. 1.16).

```
> library(hawkesbow)

> plot.hawkes.bis = function(x, intensity = FALSE,

  precision = 1e3, fun = NULL, repr = NULL,

  family = NULL, M = NULL, ...) {

  if (intensity==FALSE) {

  if (!is(x, "hawkes")) stop("'intensity==FALSE'

  is only compatible with Hawkes

 processes simulated from function 'hawkes'.")

  plot(x=NULL, xlim=c(0, x$end),

  ylim=c(-1, length(x$gen)-.5), yaxt="n",

  main = "Ostiense",

  ylab = "generations", xlab=expression(italic(t)))

  axis(2, at=1:length(x$gen)-1)

  points(x=x$gen$gen0,

  y=rep(0, length(x$gen$gen0)), pch=15)

  eps = 0.05

  if (length(x$gen) == 1) return()

  for (i in 2:length(x$gen)) {
```

```
points(x=x$gen[[i]], y=rep(i-1, length(x$gen[[i]])),
pch=19, col=i)
arrows(x0=x$gen[[i]], x1=x$gen[[i-1]][x$ancestors[[i-1]]],
y0=i-1-eps, y1=i-2+eps, code=1, length=.1)
}
points(x=x$p, y=rep(-0.5, x$n), pch=4)
segments(x0=0, x1=x$end, y0=-.5, col="grey")
}
if (intensity==TRUE) {
if (is.numeric(x)) p = x
else if (is(x, "hawkes")) p = x$p
else stop("Argument 'x' must either be
of class 'hawkes' or a numeric vector.")
if (is(x, "hawkes")) end = x$end
else end = tail(x, 1)
matplot(z = seq(0, end, by=end / precision),
zt = intensity(x = x, t = z, fun = fun, M = M,
repr = repr, family = family, ...),
type="l", ylim=c(0, max(zt)),
```

```
main = "Ostiense",

xlab=expression(italic(t)),

ylab=expression("conditional intensity"))

segments(x0=0, x1=end, y0=0, col="grey")

points(x=p, y=rep(0, length(p)), pch=4)

}
}

> x.flaminia.night = hawkes(end = 1, fun = 42.9,

repr = 0.34131, family = "exp", rate = 77.3)

> x.appia.night = hawkes(end = 1, fun = 32.3,

repr = 0.46296, family = "exp", rate = 26.7)

> p1.flaminia.night = plot.hawkes.bis(x.flaminia.night,

intensity = TRUE)

> h.flaminia = 42.9/(1-0.34131)

> abline(h = h.flaminia, col='blue', lwd=2, lty='dashed')

> p1.appia.night = plot.hawkes.bis(x.appia.night,

intensity = TRUE)

> h.appia = 32.3/(1-0.46296)

> abline(h = h.appia, col='blue', lwd=2, lty='dashed')
```

F.5. Ruin Probabilities (Chapter 5, § 5.4.1.2, Fig. 5.8 and Fig. 5.9)

The R code below computes the ruin probability of a fractional model (see chapter 5, § 5.4.1.2, Fig. 5.8 and Fig. 5.9) on the series of Italian earthquakes (from 1600).

```
> getwd()

> setwd("D:/Dati/Profili/Documenti")

> read.csv2(file = "nuovo.catalogo.20180312.csv",

   header = "TRUE")

> solve()

> root()

> my.x = seq(0,10,0.1)

> my.mu.X = 1

> my.lambda.1 = 0.43

> my.lambda.2 = 0.66

> my.nu.1 = 0.85

> my.nu.2 = 0.49

> my.fun = function(x){
  pi = 1.2
  ni = 0.49
  x - my.mu.X + my.lambda.1 * pi  (-ni) * x  (1-ni)

    }
```

```
> uniroot(my.fun, lower = 0, upper = 10)

> my.mu.X = 1

> my.x.star.1 = 0.68

> my.x.star.2 = 0.55

> psi.nu = function(u, mu.X, x.star){

    a = (1 - x.star/mu.X)

    b = exp(-x.star*u)

    c = a*b

    return(c)

    }

> my.u = seq(0, 12, 0.1)

> psi.1 = psi.nu(u = my.u, mu.X = my.mu.X,
x.star = my.x.star.1)

> head(psi.1)

> psi.2 = psi.nu(u = my.u, mu.X = my.mu.X,
x.star = my.x.star.2)

> head(psi.2)

> plot(psi.2, typ = "l", lwd = 2, col = "red",
  ylab = "psi(u)", xlab = "u", xaxt="n"
  )
```

```
> my.labels = seq(0,12,1)

> axis(side = 1, at = seq(0,120,10), labels = my.labels)

> legend(x = 95, y = 0.45,
  legend=c("v = 0.49", "v = 0.85"),
  col=c("red", "blue"), lty=1, lwd = 2
  )

> lines(psi.1, typ = "l", lwd = 2, col = "blue")
```

References

Aki S., Hirano K., Estimation of the distributions of order k, *Ann. Inst. Statist. Math., Vol. 41, No. 1, 47 61*, 1989.

Albrecher H., Asmussen S., Ruin Probabilities, *World Scientific, 2nd edition*, 2010.

Albrecht P., On the correct use of the chi-square goodness-of-fit test, *Scandinavian Actuarial Journal*, 1980:3, 149-160, 1980.

Andelson R.M., Compound Poisson distributions, *Operations Research Quarterly*, 17, 73-75, 1966.

Bailey, The Elements of Stochastic Processes with Applications to the Natural Sciences, *New York Wiley*, 1964.

Beckers S., A note on estimating the parameters of the diffusion-jump model of stock returns, *Journal of Financial and Quantitative Analysis*, 16:127–140, 1981

Beghin L., Orsingher E., Fractional Poisson Processes and Related Planar Random Motions, *Electronic Journal of Probability*, 14(61):1790-1826, 2009.

Biard, Saussereau, Fractional Poisson process: long-range dependence and applications in ruin theory, *Journal of Applied Probability* 51:727–740, 2014

Buchak K., Sakhno L., Composition of Poisson and Gamma processes, Modern Stochastics: Theory and Applications, 4(2):161-188, 2017.

Buchak K., Sakhno L., On the governing equations for Poisson and Skellam processes time-changed by inverse subordinators, *Theory of Probability and Mathematical Statistics*, 98:91-104, 2019.

Buhlman H., Mathematical methods in risk theory *Springer*, 1970 (2nd ed. 2005).

Cahoy D., Estimation of Mittag-Leffler Parameters, *Communications in Statistics–Simulation and Computation*, 42: 303–315, 2013.

Cahoy D., Uchaikin V., Woyczynski W., Parameter estimation for fractional Poisson processes, *Journal Statistical Planning Inference*, 140, 3016-3120, 2010.

Charalambides C., On discrete distributions of order k, *Annals of the Institute of Statistical Mathematics*, (38), Part A, 557-568, 1986.

Chernoff H., Lehmann E., The Use of Maximum Likelihood Estimates in 2 Tests for Goodness of Fit, *The Annals of Mathematical Statistics*, 25, 579-586, 1954.

Cinque F., Orsingher E., On the distribution of the maximum of the telegraph process, *Theory of Probability and Mathematical Statistics*, 73:95-102, 2020.

Cinque F., Orsingher E., On the exact distributions of the maximum of the asymmetric Telegraph process, *Stochastic Processes and their Applications*, 142:601-633, 2021.

Clarke R.D., An application of the Poisson distribution, *Journal of the Institute of Actuaries*, Vol. 72, Issue 3, p. 481, 1946.

Constantinescu C., Ramirez J., Zhu W., An application of fractional differential equations to risk theory, *Finance and Stochastics*, 23:1001-1024, 2019.

Constantinescu C., Ni W., Pantaleous A., Bonus–Malus systems with Weibull distributed claim severities, *Annals of Actuarial Science*, 8(2):217-233, 2014.

Cox D.R., Some statistical methods connected with series of events, *Journal of the Royal Statistical Society*, B, 17:129-164, 1955.

Cox D.R., Lewis P.A.W., The statistical analysis of series of events, *John Wiley and Sons*, New York, USA, 1966.

De Moivre A., De mensura sortis, seu, de probabilitate eventuum in ludis a casu fortuito pendentibus, *Philosophical Transactions of the Royal Statistical Society*, 329(1):213-264, 1711.

Davison A.C., Hautphenne S. and Kraus A., Parameter estimation of discretely observed linear birth-and-death processes, *Biometrics*, 77:186-196, 2021.

Di Crescenzo A., Logistic Growth described by Birth-Death and Diffusion Processes, *Mathematics*, 7:489, 2019.

Duffie D., Sigleton K., Credit Risk: Pricing, Measurement, and Management, *Princeton University Press*, 2003.

Feller W., On a general class of "contagious" distributions, *Annals of Mathematical Statistics*, 14(4):389-400, 1943.

Frangos N., Vrontos S., Design of optimal Bonus-Malus systems with a frequency and a severity component on an individual basis in automobile, insurance, ASTIN Bulletin, 31:1–22, 2001.

Furry W.H., On fluctuation phenomena in the passage of high-energy electrons through lead, *Physical Review*, 52:569–581, 1937.

Gallager R., Discrete Stochastic Processes, *Kluwer*, 1996.

Garra R., Orsingher E., Scavino M., Some probabilistic properties of fractional point processes, *Stochastic Analysis and Applications*, 35(4):701-718, 2017.

Gurland J., Some Applications of the Negative Binomial and Other Distributions, *American Journal of Publich Health*, 49:1388-1399, 1959.

Kataria K., Khandakar M., Time-changed space-time fractional Poisson process, *Stochastic Analysis and Applications*, 2021.

Keiding N., Estimation in the birth process, Biometrika, 61 (1):71, 1974.

Keiding N., Maximum Likelihood Estimation in the Birth and Death Process, *The Annals of Statistics*, 3, No. 2, pp. 363-372, 1975.

Keiding N., Beyer J., Simonsen W., The Exact Behaviour of the Maximum Likelihood Estimator in the Pure Birth Process and the Pure Death Process, Scandinavian Journal of Statistics, 3:61-72, 1976.

Kendall D., The generalized birth-and-death process, *Annals of Mathematical Statistics*, 19, pp. 1-15, 1948.

Kendall D., Stochastic processes and population growth, Journal of the Royal Statistical Society. Series b (Methodological), Vol. 11, No. 2, 1949.

Kirschenhofer P., A note on alternating sums, *The Electronic Journal of Combinatorics*, 3(2), 1996.

Kostadinova K., Minkova L., On the Poisson process of order k, *Pliska Studia Mathematica Bulgarika*, 2013.

Lando D., Credit Risk Modeling, *Princeton University Press*, 2004.

Laub J., Lee Y. and Taimre T., The Elements of Hawkes Processes, *Springer*, 2022.

Lemaire J., Bonus-Malus Systems in Automobile Insurance, *Kluwer Academic Publisher*, 1995.

Maheshwari A., Orsingher E., Superposition of time-changed Poisson processes and their hitting times, 2019.

Meerschaert M., Nane E., Vellaisamy P, The Fractional Poisson Process and the Inverse Stable Subordinator, *Electron. J. Probab.* 16:1600-1620, 2011.

Orsingher, Battaglia, Probability distributions and level crossings of shot noise models, *Stochastics*, 8:45-61, 1982.

Orsingher E., Polito F., Fractional pure-birth processes, *Bernoulli*, 16(3):858-881, 2010.

Orsingher E., Polito F., On the fractional linear birth-death process, *Bernoulli*, 17(1):114-137, 2011.

Orsingher E., Polito F., Compositions, Random Sums and Continued Random Fractions of Poisson and Fractional Poisson Processes, *Journal of Statistical Physics*, 148(2):233-249, 2012 (a).

Orsingher E., Polito F., The space fractional Poisson processes, *Statistics and Probability Letters*, 82(4):852-858, 2012 (b).

Orsingher E., Polito F., On the integral of fractional Poisson processes, *Statistics and Probability Letters*, 83(4):1006-1017, 2013.

Orsingher E., Polito F., Sakhno L., Fractional Nonlinear, Linear and Sublinear Death Processes, *Journal of Statistical Physics*, 141(1):68-93, 2010.

Orsingher E., Toaldo B., Counting processes with Bernstein intensities and random jumps, *J. of Applied Probability*, 52(4):1028-1044, 2015.

Pearson K., On the criterion that a given system of deviations from the probable in the case of a correlated system of variables is such that it can be reasonably supposed to have arisen from random sampling, *Phil. Mag.*, 5 (50), pp. 157-175, 1900.

Philippou A., Poisson and compound Poisson distributions of order k and some of their properties, *J. Soviet Maths*, 1984.

Plackett R.L., Karl Pearson and the Chi-Squared Test, International Statistical Review / Revue Internationale de Statistique, Vol. 51, No. 1, pp. 59-72, 1983.

Poisson, S.-D., Recherches sur la probabilité des jugements en matière criminelle et en matière civile, précédées des règles générales du calcul des probabilités, *Bachelier*, 1837.

Pollard R., Benjamin D. and Reep C., Sport and the Negative Binomial Distribution, in Ladany S., Machol R. (eds.), *Optimal Strategies in Sports*, 1977.

Press S. J., SA compound events model for security prices, *Journal of Business*, 40, 317–335, 1967.

Rang G., Shen Z., Zuo H., Hitting probabilities of weighted Poisson processes with different intensities and their subordinations, *Acta Mathematica Scientia*, 41B(1), 67-84, 2021.

Reep C., Benjamin D. and Pollard R., Skill and Chance in Ball Games, *Journal of the Royal Statistical Society*, A, 134, 623-629, 1971.

Richardson L., The Distributions of Wars in Time, Journal of the Royal Statistical Society, A, 107, 242-250, 1945.

Rolski T., Schmidli H., Schmidt V., Teugels J., Stochastic Processes for Insurance and Finance, *John Wiley and Sons*, 1999.

Ross, S.M., Introduction to Probability Models, *Academic Press*, 13th Ed., 2024.

Schilling, R.I., Song R., Vondracek Z., Bernstein Functions. Theory and Applications, DE GRUYTER, 2012

Schottky W., Theory of shot effect, Ann. Phys., 57, 541-568, 1918.

Shreve, S.E., Stochastic Calculus for Finance II, Continuous time models, Springer, 2004.

Tavaré S., The linear birth–death process: an inferential retrospective, *Advances in Applied Probability*, Vol. 50, pp. 253-269, 2018.

Waugh, W., Conditioned Markov processes, *Biometrika*, Vol. 55, No. 1/2, pp. 241-249, 1958.

Yule, G.U., A mathematical theory of evolution based upon the conclusions of Dr. J.C. Willis, *Phil. Trans. Roy. Soc. Lond. B*, 213, pp. 21-87, 1924.

Von Bortkewitsch L., Das Gesetz der kleinen Zahlen, *Teubler*, 1898.

Analytic Index